Proceedings of the Institute of General Physics
Academy of Sciences of the USSR
Series Editor: A.M. Prokhorov

SELECTIVE LASER SPECTROSCOPY
ACTIVATED CRYSTALS AND GLASS

Proceedings of the Institute of General Physics
Academy of Sciences of the USSR
Series Editor: A.M. Prokhorov
Volume 9

SELECTIVE LASER SPECTROSCOPY OF ACTIVATED CRYSTALS AND GLASSES

Edited by V.V. Osiko
Translated by Kevin S. Hendzel

Nova Science Publishers
New York

Deputy Series Editor: T.B. Voliak

Nova Science Publishers, Inc.
283 Commack Road
Suite 300
Commack, New York 11725

This book is being published under exclusive English language rights granted to Nova Science Publishers, Inc. by the All-Union Copyright Agency of the USSR (VAAP).

Library of Congress Cataloging-in-Publication Data available upon request

ISBN   0-941743-92-6

The original Russian-language version of this book was published by Nauka Publishing House in 1988

Graphic Design by Elenor Kallberg and Peggy Harvey

TABLE OF CONTENTS

FROM THE EDITOR

The extensive applications of lasers in scientific research has given rise to rapid development of such fields of science as photochemistry, photobiology, nonlinear optics, and laser spectroscopy of gases and solids. Researchers in solid state physics have especially focused on the structurally-disordered state of matter. Interest in such objects (crystalline solids, glasses, and liquids) can be attributed to their extraordinary variety as well as the uniqueness of their physical properties, which has given rise to new prospects for their practical application.

Inorganic crystals and glasses activated by rare earth elements and Fe-group elements are of enormous practical value as laser and luminescent media for quantum electronics and optics. These also represent excellent model objects for investigating local electron and vibronic states, ion-ion interaction and nonradiative electron excitation energy transfer in solids.

The substantial dispersion of both the energy and relaxation properties of active ions in disordered matter is responsible for the complex inhomogeneous nature of its optical spectra and the deactivation kinetics of optical excitations which in turn complicates the application of traditional spectroscopic techniques to investigations of such objects. In this regard the development of new selective nonstationary laser luminescent spectroscopic techniques that reveal the nature of the inhomogeneous optical characteristics of such media and can be used to investigate their properties on the microlevel assumes an important role.

The first sample application of spectral selection of centers in disordered condensed matter was in the study by Yu. V. Denisov and V. A. Kizel' (1967) which analyzed europium ions in glass. A mercury lamp was used in this study as the monochromatic source. This method was then used with lasers to investigate glasses containing ytterbium in the study by I. V. Vasil'ev et al. (1969). The selective laser excitation technique was then used successfully to analyze ruby crystals in Szabo's study (1970), for frozen organic solutions in a study by V. I. Personov and his associates (1972) and to investigate neodymium glass in the study by L. A. Riseberg (1972).

While recognizing the pioneering and fundamental nature of such research it is important to note a significant deficiency in the formulation of the experiments; specifically, the use of fixed-frequency monochromatic light, which complicated the detection and analysis of the varied, complex electron structure of the collection of local optical centers in disordered matter. This limitation was eliminated in the study by N. Montegi and S. Shionoya (1973) in which the authors employed a tunable rhodamine 6Z laser thereby making it possible to identify the complex, quasicontinuous nature of optical europium center formation in phosphate glass and the monotonic correlated nature of the variation in the energies of the Stark levels of this ion.

The staff of the Division of Solid State Physics of the Institute of General Physics of the USSR Academy of Sciences developed an original pulsed laser-luminescent spectrometer in 1974. This instrument included a tunable nanosecond rhodamine 6Z dye laser producing a pulse power of 10^5 W operating over a tuning range of 0.55-0.63 μm with a lasing spectral width of less than 1 Å. The laser operated in a pulse-periodic mode and was used for "instantaneous" frequency-selective excitation of individual groups of activated europium, samarium and neodymium centers in crystalline solid solutions and glasses.

Nonstationary luminescence was recorded by automatic individual photon counting using a high-sensitivity delayed coincidence method (sensitivity on the order of tens or hundreds of photons per second). This instrument provided the central element of the laser spectroscopic system and made it possible for the staff at the Institute of General Physics to be among the first in the nation to develop and apply instantaneous selective laser excitation and stroboscopic photographic recording techniques to a systematic investigation of the spectra of disordered crystals and glasses containing rare earth (RE) ions.

The establishment of the new field of the physics of activated crystals and glasses gave rise to a broad range of engineering, methodological, and fundamental physical problems for consideration by researchers. These included the development of continuously-tunable laser light sources operating in the required wavelength range, pulsed photographic recorders, selective laser spectroscopy schemes and techniques, and the use of techniques for analyzing the inhomogeneously-broadened spectra and relaxation characteristics of the elementary optical centers of rare-earth ions in disordered matter together with technique selection and direct investigation of the quenching kinetics and energy migration processes in active ions in disordered solids.

The solutions to these problems led to the development and implementation of selective spectroscopy techniques based on measurement of nonstationary luminescence and excitation spectra together with the damping kinetics in conditions of selective excitation and recording. A number of interesting results were obtained from using such research techniques. We will outline the fundamental results.

The previously-unknown structure of the electron and vibronic states was identified for a number of crystalline and glass-like media activated by europium, samarium, neodymium, and ytterbium ions with inhomogeneously-broadened spectra, together with the magnitude of homogeneous spectral broadening together with the intracenter energy relaxation channels. A regular rise in the symmetry of the coordination polyhedron of the impurity ion with diminishing ion-ligand distance was established based on neodymium, ytterbium, and chromium ions in various glasses. A spectral-kinetic method of analyzing the spatial distribution of active ions in disordered matter was proposed and implemented. A dipole-quadrupole mechanism was detected and both the macro- and micro-parameters of quenching samarium ion interaction in the glass were determined. Detection of the initial ordered stage of quenching kinetics in glass is of substantial interest for developing the concepts relating to the electron excitation relaxation mechanism in disordered matter. A short-range order parameter — the minimum distance — is determined; this characterizes the relative location of the impurity ions in the disordered medium.

A new method of investigating the frequency dependence of elementary, nonresonant interactions was proposed and substantiated theoretically; this technique is based on an analysis of the low-temperature selective luminescence spectra. The exponential dependence of the microefficiency of nonresonant energy transfer on the frequency of the emitted phonon is noted.

Multistage resonant energy migration among an inhomogeneous particle ensemble due to the overlapping of their homogeneous spectra and characterized by an increase in the Lorentz broadening of the narrow spectral component with increasing delay is detected and investigated.

An exponential ordered migration stage was discovered experimentally for the first time for activated glasses together with a disordered "Foerster" excitation migration stage. A quadrupole-quadrupole reversible interaction mechanism of the europium ions in glass was established. The migration rates were determined together with the effectiveness of elementary migrational europium and ytterbium ion interaction and the distance of closest approach of the europium ions in silicate glass.

The existence of a diffuse, stationary electron excitation migration stage in a disordered particle group was identified experimentally for the first time. The onset time of this migration, the hyperbolic (power of 3/2) kinetic dependence together with the disordered diffusion constants and their concentration relations are determined.

All these results relating to the development and application of nonstationary selective laser spectroscopy techniques are presented in the present volume and are important for the development of the physics of activated glasses and crystals with a disordered structure.

SPECTRAL AND RELAXATION CHARACTERISTICS OF LOCAL ELECTRON STATES OF IMPURITIES IN STRUCTURALLY-DISORDERED MATRICES

O. K. Alimov, T. T. Basiev, S. B. Mirov

Abstract: The significant capabilities of kinetic selective laser excitation luminescent spectroscopy are demonstrated based on an investigation of crystals and glasses with Nd^{3+}, Eu^{3+}, Sm^{3+}, Yb^{3+}, and Cr^{3+} ions. The present article discusses both existing and new selective spectroscopy schemes for analyzing electron and vibronic states, radiative and nonradiative relaxation channels, and the ion-lattice interaction mechanisms in disordered condensed matter with inhomogeneous spectral broadening of the rare-earth ions.

Selective laser spectroscopy is employed to identify the structure of inhomogeneous broadening, while the details of Stark splitting of the inhomogeneously-broadened absorption and luminescence spectra of the ions outlined above in a number of crystalline and glass-like matrices are detected and investigated. General distribution maps of the Stark sublevels are plotted for continuous sets of optical centers in glass. The causes of inhomogeneous spectral broadening of a number of electron transitions are discussed.

The temperature dependences of homogeneous broadening of the inhomogeneously-broadened spectra of Yb^{3+}, Nd^{3+}, Sm^{3+}, and Eu^{3+} in glasses and crystals are measured; these reveal a relaxational broadening mechanism due to single-phonon inter-Stark nonradiative transitions.

The vibronic spectra of Yb^{3+} in Ba-Al-phosphate glass was identified, thereby demonstrating the exponential frequency dependence of the vibrational state density in the range 20-200 cm^{-1}. The quasihomogeneous and inhomogeneous broadening of the vibronic luminescence spectra of F_2^+-color centers in LiF crystals are estimated.

Nonstationary luminescence and excitation spectroscopy techniques reveal a substantial dispersion of the level lifetimes together with the intra-center radiative (in the case of Nd^{3+}, Yb^{3+}, Cr^{3+}, and Eu^{3+} ions) and nonradiative (for Cr^{3+}) transition probabilities for a number of glasses; this correlates with the changing energy of the electron transitions and Stark splittings from one optical center to another. The mechanism of increasing symmetry of the coordination polyhedron of the impurity atom with diminishing ion-ligand distance is established based on sample Nd^{3+}, Yb^{3+}, and Cr^{3+} ions in various glasses.

Laser luminescence excitation sources used in conjunction with stroboscopic photographic recorders which fix radiation (record spectra) at arbitrarily-selected times following excitation can be used to approach the study of the Stark and vibronic structure of the inhomogeneously-broadened spectra of rare-earth (RE) ions and relaxation processes concealed by inhomogeneous broadening on a qualitatively new level.

Inhomogeneous spectral broadening is the manifestation of the structural disorder of an activated medium. In such promising media for solid state physics, quantum electrons, and optics as compound crystals and glasses [1-4] the inhomogeneous spectral broadening of the rare-earth ions is so substantial that even at high temperatures (300 K) it partially or entirely moderates the spectral properties of the individual optical centers: Stark splitting, the radiative transition probabilities, the lifetimes and magnitude of homogeneous broadening.

Such a role of inhomogeneous broadening begins to appear for media with a high degree of structural perfection (crystals) only at lower temperatures ($T \gtrsim 77$ K) when homogeneous broadening is substantially reduced.

The lack of information on the structure of the elementary components of the inhomogeneously-broadened bands and the energy exchange rates within the inhomogeneous contour substantially complicates the analysis of processes occurring in an excited, activated medium, making it impossible to produce an unambiguous prediction of its behavior in quantum instruments. The selective laser excitation (SLE) method may fill this gap.

1. Selective Methods in the Spectroscopy of Activated Media

The selective laser excitation method is based on narrowing of inhomogeneously-broadened luminescence lines by narrowband selective excitation of individual groups of centers (spectral packets) from an inhomogeneous ensemble. In this method it is necessary to differentiate the two luminescence observation schemes: the resonant measurement scheme and the nonresonant (relative to the transition where selective ion excitation occurs) scheme. When luminescence is recorded at the same transition where selective excitation is initiated (the resonant scheme) a greater narrowing of the luminescent lines is observed, since the inhomogeneous component of luminescence line broadening in this case diminishes to the laser linewidth or to the homogeneous width of the transition and may be quite small.

Inhomogeneously-broadened spectral narrowing is also observed in the nonresonant observation of luminescence, although it may be less clearly expressed due to the incomplete correlation of the energies of the radiative transition and the transition at which excitation occurs.

Reducing the inhomogeneous component of spectral line broadening by the selective laser excitation method makes it possible to identify the following elementary characteristics of individual groups of optical centers in a disordered medium:

1) the magnitude of homogeneous broadening of spectral lines, their variation from center to center in the glass and the temperature relations;

2) the Stark splitting energy at various optical centers, the dispersion of such Stark splitting events, and the correlation of the various electron transition energies.;

3) the luminescence decay kinetics of the individual groups of centers, their lifetimes, as well as the radiative and nonradiative transition probabilities and their dispersion;

4) the elementary interaction probabilities of laser-excited centers with spectrally-different, nonexcited centers (spectral energy migration).

The resonant selective excitation technique was first applied to the investigation of inhomogeneously-broadened spectra and energy migration across the spectra in Refs. [5-7] by investigating the $^5D_0-^7F_0$, $^7F_1$ transitions of Eu^{3+} ions in sodium-borate glasses. The authors employed the narrow spectral lines of a mercury lamp — $\lambda_1 = 576.9$ nm and $\lambda_2 = 579.1$ nm — as the monochromatic radiation source. In spite of a number of inaccuracies in the spectral interpretation, such a selection made it possible for the authors to observe, for the first time, a narrowing of the inhomogeneously-broadened spectra and to employ the temperature and concentration variations of such spectra to analyze the energy migration process under near-steady-state conditions.

Using the monochromatic radiation from a tungsten lamp for selective excitation of Yb^{3+} ions in glass made it possible for the authors of Ref. [8] to identify the spectral features of low-temperature excitation migration among an inhomogeneous ensemble of centers, thereby emphasizing the critical role of transfer processes involving emission of phonons.

A highly monochromatic ruby laser ($\Delta\nu_L = 150$ MHz) was subsequently used by the authors of Ref [9, 10] to implement selective excitation of Cr^{3+} ions in ruby at the $^4A_2-^2E$ transition which has a narrow, inhomogeneous width ($\Delta = 2.2-4.55$ GHz) and to investigate its Zeeman structure.

CW argon laser radiation at $\lambda = 472.7$ nm in a nonresonant scheme and operating at the $^4I_{9/2}-^4G_{11/2}$ transition was used in 1972 by L. A. Riseberg et al. [11-13] to excite Nd^{3+} ions in silicate glass. In steady-state conditions Riseberg observed narrowing of the short wavelength components of the luminescence lines at the $^4F_{3/2}-^4I_{9/2}$ and $^4F_{3/2}-^4I_{11/2}$ transitions, indicating a correlation of their energies. Riseberg treated the concentration and temperature variations of the spectrum in the language of spectral energy migration among the $^4F_{3/2}$ metastable levels of the Nd^{3+} ions.

In this same year the authors of Refs. [14, 15] demonstrated the promise of using selective laser excitation to investigate organic glasses (frozen solutions). An He-Cd laser was used to identify for the first time the zero-phonon line and the vibronic structure of the spectrum for frozen ethanol glasses with substantial inhomogeneous broadening.

We note that the development and application of selective laser spectroscopy techniques to two important classes of disordered solids — organic molecular crystals and glasses and inorganic crystals and glasses with active ions — largely occurred independently. This was most likely due to the broad variety of objects as well as the substantial differences in their optical properties such as the transition type and probability, lifetime, electron-phonon interaction, and photochemical stability of the optical centers.

In addition to general interest in studies of homogeneous broadening and vibronic interaction, these features identified a number of effects for use in specific investigations of molecular systems such as the Zeeman effect, null-field line splitting, the spectroscopy of biological objects, photochemical reactions in solids and spectrochemical analyses of mixtures. Although here we provide only a list of these important problems, we cite survey studies [16-18] where a detailed analysis and extensive references on these issues can be found.

The selective luminescence spectroscopic techniques for application to disordered media under monochromatic excitation which are the focus of the present study are closely related to the monochromatic-induced-relaxation [9-28] and photochemical hole "burning" [29-31] techniques in inhomogeneously-broadened luminescence lines as well as techniques used to investigate the polarization luminescence of the inhomogeneously-broadened bands of rare earth ions under excitation by linearly-polarized light [32-34].

We will examine in greater detail the development and application of selective laser spectroscopy techniques to a class of impurity inorganic solids.

The spectral analysis of the $^4F_{3/2}-^4I_{11/2}$ transition of Nd^{3+} ions in $CaF_2-YF_3 : Nd^{3+}$ and $SrMoO_4 : Nd^{3+}$ crystals carried out by the authors of Ref [35] by excitation at the $^4I_{9/4}-^4F_{3/2}$ transition using a monopulse liquid tunable laser revealed substantial distortions to the luminescence contour (excitation selectivity) compared to the contour with nonselective excitation. In spite of insufficient spectral resolution, the authors were able to identify several specific types of centers (at least two for $CaF_2-YF_3 : Nd^{3+}$) in these matrices.

One of the studies that has demonstrated the extensive capabilities and substantial information content of selective spectroscopy using continuously tunable lasers is Ref. [36] devoted to an analysis of Eu^{3+} ions in $Ca(PO_3)_2$ glass. The authors identified the complex structure of inhomogeneous broadening in glass, the dispersion and correlation of Stark splittings of the $^7F_1$ and $^7F_0$ levels and analyzed energy migration in the nonstationary luminescence spectra.

The extensive use of rhodamine 6Z dye lasers which have a rather broad operational tuning band (550-630 nm) and a high degree of monochromaticity has made it possible to investigate the inhomogeneously-broadened spectra of Eu^{3+}, Pr^{3+}, and Sm^{3+} ions whose metastable levels lie in the photon energy range of this laser [36-49]. The Stark structure of the inhomogeneously-broadened bands of Eu^{3+} in Ca-Eu-phosphate and Eu-metaphosphate glasses was investigated by two teams of authors [36, 37] in conditions of both resonant and nonresonant luminescence excitation.

Ref. [38] used an analogous monochromatic excitation method at the $^7F_0-^5D_0$ transition to investigate the Stark structure of the inhomogeneously-broadened spectra of $^5D_0-^7F_1$ and $^5D_0-^7F_0$ as well as variations in the luminescence quenching time of the Eu^{3+} ions in Ba-Na-Zn-silicate glass. The luminescence spectra were calculated by crystal field theory within the framework of C_{2v} symmetry and such calculations were used as the basis for proposing a nearest oxygen neighbor model of the Eu^{3+} ions in the glass matrix.

This same matrix was used in Ref [39] to obtain the dependence of relaxational broadening at various centers on the energy gap ΔE between $^7F_0$ and the lower component of the $^7F_1$-level. This study reports observing variations in the lifetimes of the Nd^{3+} and Yb^{3+} ions at different optical centers of the glass and also investigated spectral migration at the $^2F_{5/2}-^2F_{7/2}$ transition of the Yb^{3+} ions.

The use of resonant luminescence excitation at the $^1D_2-^3H_4$ transition of Pr^{3+} ions makes it possible to reduce the inhomogeneous component of line broadening to the linewidth of the exciting laser; this inhomogeneous component is easily narrowed to 0.03-0.08 cm^{-1} for a rhodamine 6Z laser. This technique was used to investigate distortions to the symmetry of the Pr^{3+} ions in various centers in an $LaAlO_3$ lattice [40] and for investigating the temperature relations for homogeneous broadening of the $^1D_2-^3H_4$ transition of Pr^{3+} ions in LaF_3 in the low temperature range (10-50 K) when the homogeneous width is substantially less than the inhomogeneous width [41]. Refs [42, 43] are devoted to an investigation of energy migration over the metastable $^3P_0$ level of Pr^{3+} ions in an LaF_3 matrix; this study employs concentration, temperature, and temporal relations of luminescence in the case of selective excitation to identify the dipole-dipole nature of $Pr^{3+}-Pr^{3+}$ phonon-assisted interaction.

Ref. [44] employed laser spectroscopy to obtain the spectra of several optical centers of Sm^{3+} in a $CaWO_4$ crystal. Temperature broadening and spectral line shifts were investigated. Energy migration among the Sm^{3+} ions was investigated as a function of temperature, concentration, and pressure.

The homogeneous broadening parameters of the inhomogeneously-broadened lines of Eu^{3+} in glasses and crystals were measured in Refs. [45-48]; these studies compared and analyzed the magnitude of broadening in a number of matrices with a similar nearest neighbor structure of the active ion. Ref. [49] first detected the polarization memory of the luminescence spectra in glass using Ca-phosphate glass with Eu^{3+} for the case of selective excitation of the Eu^{3+} ions by polarized laser radiation.

An analysis of the literature data suggests that the greatest increase in the number of publications and the subsequent sustained interest in the use of spectrally-selective methods of investigating activated media occurred in the period 1975-1976. This period also represents the end of the development of techniques begun in 1974 and the initial investigations of selective laser spectroscopy of activated media carried out by the staff of the Institute of General Physics of the USSR Academy of Sciences [50-55].

Surveys of the most recent studies devoted to the methods, techniques, and application of selective spectroscopy of activated crystals and gases which, specifically, identify the contribution of the staff of the Institute of General Physics to the development of this field, can be found in Refs. [33, 56-60].

2. Laser-Luminescent Spectrometer for Nonstationary Selective Spectroscopy of Activated Solids

The studies of the electron structure and relaxation properties of rare earth activators, their ion-to-ion interactions and electron excitation migration for disordered condensed matter with inhomogeneous spectral broadening presented in the present study have determined the specific requirements that must be satisfied by the laser spectroscopic instrument. Its primary operational measurement modes include the following:

1) measurement of the nonstationary luminescence spectra $I(E_L)$ for discrete monochromatic excitation frequencies[1] E_L at a high temporal resolution ($\Delta t \rightarrow 0$) and fixed delays between the excitation time and the recording time t_d;

2) measurement of the luminescence decay kinetics $I(t)$ for discrete monochromatic excitation frequencies E_L and luminescence recording frequencies E_R;

---

[1] Henceforth we will not differentiate between the frequency and energy of radiation which are unambiguously related by the relation $E = h\nu$ and expressed in cm^{-1}

3) measurement of the nonstationary excitation spectra $I(E_L)$ for discrete luminescence recording frequency E_R and delays t_d.

The extensive variety of electron levels of rare earth ions and the enormous number of absorption and luminescence transitions extending over a broad spectral range from the ultraviolet through the infrared require the development of broadband luminescence photographic recorders and monochromatic excitation sources.

The optical transition rates and lifetimes of the metastable levels of rare earth ions in solids ($\tau = 10^{-5}$-10^{-3} s) determine the temporal range for investigating the luminescence decay kinetics and the strobing time scale of nonstationary illumination (10^{-6}-10^{-3} s).

Photoelectron multipliers are used for photographic recording of luminescence; these devices are used successfully in both an analog recording mode (for the average signal intensities) and in a photon counting mode for counting individual photons (for extremely weak luminous fluxes). These devices satisfy the high sensitivity requirements and also have a temporal resolution sufficient for recording the luminescence of rare earth ions.

The best source for "instantaneous" ($t_e \ll \tau$) monochromatic excitation of rare earth ions includes nanosecond tunable organic dye and color center crystal lasers operating at pulse repetition rates of $f = 10$-100 Hz ($f \ll \tau^{-1}$). The rather high output radiation intensity of such lasers, running to 10^{15}-10^{17} photons per pulse, makes it possible even with inefficient utilization (to excite rare earth ions with substantial quenching or operating at transitions with very low optical density) to measure luminescence in the latter stimulation stages after it has decayed by several orders of magnitude [50, 51]. The scale of average Stark splittings of the multiplets of rare earth ions amounting to $\Delta E \approx 100$-2000 cm^{-1} and their substantial dispersions (ΔE_i) from center-to-center in disordered matrices results in inhomogeneous broadening $\Delta \sim \Delta E$. This is responsible for the requirement for a maximum continuous tunable laser scanning range in order to measure the laser excitation spectra to identify the fine structure of the inhomogeneous bands ($E_L^{max} - E_L^{min} \gtrsim 10^3$ cm^{-1}). Here the maximum permitted lasing band of the tunable laser is determined by the magnitude of homogeneous broadening of the transitions of the rare earth ions which ranges from 0.3 to 50 cm^{-1} in the temperature range 77-300 K.

We took these requirements into account in developing the laser-luminescent spectroscopic system for investigating activated solids whose block diagram is shown in Figure 1. The pump source for the two continuously-tunable laser groups was the fundamental and second harmonic from a nanosecond pulsed periodic neodymium + garnet laser. This pump source was selected due to its high reliability and the extensive availability of such lasers. With a high output power in the near infrared and visible ranges such lasers can easily be modified to obtain both microsecond and picosecond oscillation pulses.

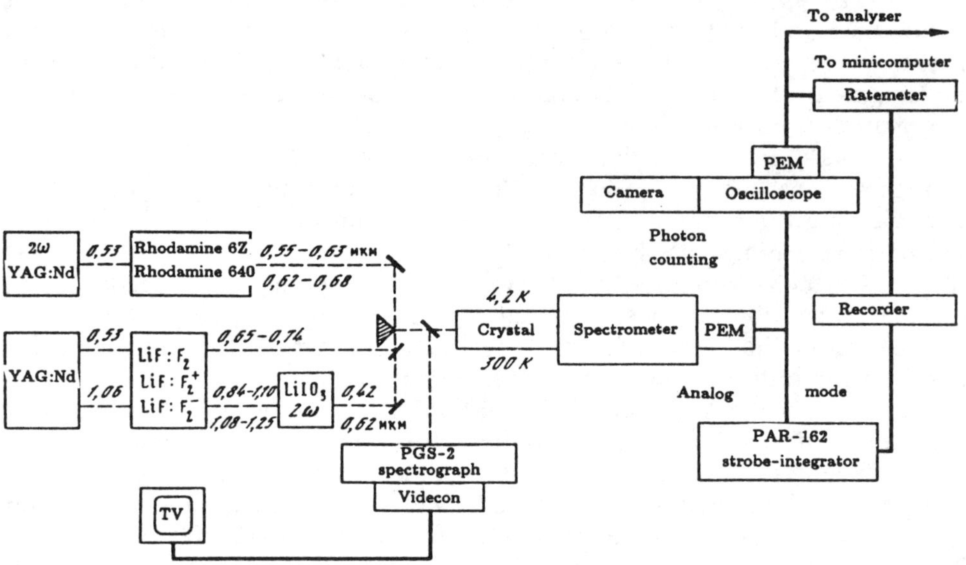

Fig. 1. Block diagram of system for nonstationary selective laser excitation spectroscopy.

The first group of lasers includes a tunable laser with two replace-able dyes: rhodamine 6Z and rhodamine 640. By using slit and holographic reflection gratings this laser can cover the 0.55-0.68 μm range at a linewidth of less than 1 Å. The second group includes solid state tunable lasers operating at the F_2^-, F_2^+, and F_2^--color centers (CC) in an LiF crystal. Such lasers have a high pump radiation-to-lasing conversion efficiency amounting to tens of percent and operate at room temperature with a high pulse repetition rate (above 50 Hz) without forced cooling. These lasers produce narrowband radiation ($\Delta\lambda \leqslant 1.0$ Å) which is continuously tunable over a broad range of wavelengths: 0.65-0.74, 0.84-1.1, and 1.08-1.25 μm, where dye lasers are less effective.

The development of tunable lasers employing the F_2^-, F_2^+, and F_2^--color centers and a single pump source (a neodymium laser) makes it possi-ble to provide nearly continuous coverage of the 0.65-1.25 μm spectral range which has important practical applications and is poorly handled by dye lasers. Figure 2 shows the tuning ranges of the LiF color center lasers and the dye lasers. It is clear that the F_2^-, F_2^+, and F_2^--color centers in LiF can replace at least eight different dyes, as they exceed these dyes in continuous scanning coverage and efficiency and eliminate the need to use a large number of complex and cumbersome flow systems. Frequency doubling of the color center lasers makes it possible to advance into the blue-green (0.42-0.55 μm) and further into the yellow and red regions (0.54-0.62 μm) of the optical spectrum and to provide nearly complete coverage of a broad wavelength range.

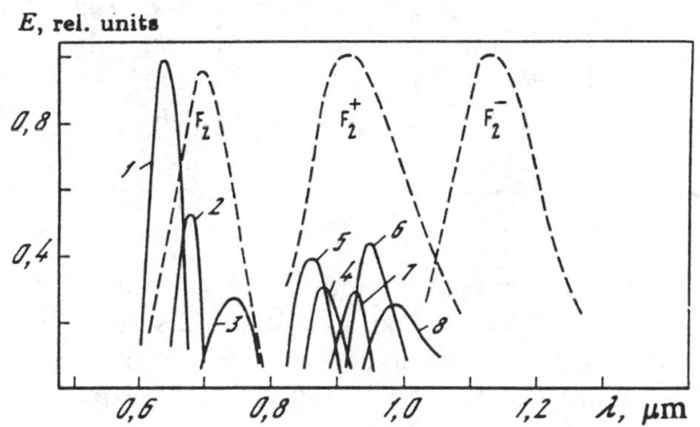

Fig. 2. Tuning ranges of F_2, F_2^+, and F_2^- LiF color center and dye lasers. 1 - rhodamine B; 2 - rhodamine 640; 3 - oxazine 720; 4 - PK 830; 5 - PK 788; 6 - PK 865; 7 - PK 850; 8 - PK 890.

Table 1 provides a number of nonlinear frequency conversions for color center lasers that we have implemented by means of $LiIO_3$, $LiNbO_3$, and GaSe crystals. By mixing the high-power radiation at the fundamental of the neodymium pump laser v_1 with the radiation from tunable lasers v_i in a nonlinear crystal it was possible to substantially enhance the conversion efficiency to the blue-green portion of the spectrum ($v = v_i + v_1$) and to advance into the far infrared ($v = v_i - v_1$) range of 5.5-16.5 μm. We note that the scheme for generating difference frequencies $v = v_i - v_1$ using an LiF laser ($F_2 \rightarrow F_2^+$) proposed in Ref. [61] yields a very broad range of continuously tunable radiation employing a single GaSe crystal and has no other constraints in the range of long oscillation wavelengths aside from the vibrational optical absorption range for the given nonlinear crystal.

The radiation wavelength and lasing line shape is monitored by a television receiver in the spectroscopic system using a PGS-7 spectrograph equipped with a television camera [62]. The laser-excited luminescence from the crystal is recorded after the monochromator (DFS-12, SDL-1, MDR-2) by a photomultiplier (FEU-79, FEU-70, FEU-62, FEU-83) in one of two modes: an analog mode or a counting mode. Both of these modes make it possible to implement temporal strobing of the luminescence signals either in sync or with a fixed delay t_d relative to the laser excitation pulse.

Table 1

Spectral lasing ranges λ of color center lasers and their nonlinear conversion

Laser type	Frequency	λ, μm	Type of nonlinear conversion	Nonlinear crystal	η, μm
YAG:Nd	ν_1	1.06	$2\nu_2$	LiIO$_3$, LiNbO$_3$	0.54–0.615
LiF : F$_2^-$	ν_2	1.08–1.23	$\nu_2 + \nu_1$	LiIO$_3$, LiNbO$_3$	0.535–0.57
LiF : F$_2^+$	ν_3	0.84–1.1	$2\nu_3$	LiIO$_3$	0.42 –0.55
LiF : F$_2$	ν_4	0.65–0.75	$\nu_3 + \nu_1$	LiIO$_3$	0.47 –0.54
			$\nu_3 - \nu_1$	GaSe	5.5 –16.5

In the analog measurement mode the photomultiplier signal is injected to the input of a PAR-162 single-channel strobe-integrator which is in sync with the laser excitation pulses. The strobe-integrator makes it possible to measure and average the instantaneous luminescence signal values over many pulses at any point on the kinetic excitation decay curve $I(t_d)$ with a variable delay t_d and a time window Δt selected in the range from 10^{-8} to 10^{-2} s.

The oscillographic delayed coincidence method proposed in Ref. [63] and developed in Refs. [64, 65] was used in the photon counting mode to implement time strobing of the luminescence signal. This technique is simple and permits accessibility to the instrumentation, while providing high sensitivity, noise immunity, and clarity. This photon counting method makes it possible to measure exceedingly weak luminous fluxes on the order of hundreds and thousands of quanta per second. Its temporal resolution (t_d and Δt) is 10^{-6}—10^{-3} s, and when necessary can be improved by more than two orders of magnitude [64, 65]. The dynamic range of the method is at least three orders of magnitude.

Both photographic recorder operating modes — analog and counting — can be used to measure the averaged luminescence signal as a function of three continuously- or discretely-variable parameters: the delay t_d, the laser excitation frequency E_L and the luminescence recording frequency E_R. Final signal recording is implemented by recorders (a PDS-021 or an H306) on paper or can be input to the memory of the digital oscilloscope by using a PAR-4202 signal averager. This averager can also be used for direct multichannel (2 × 1024 or 1 × 2048 channels) recording of the luminescence decay kinetics by measuring the photomultiplier signal in real time at a sampling rate of greater than 5 μs/channel, followed by storage and averaging. The system also can output data to a multichannel analyzer (an NTA-512 or ORTEK-7032) and an LSI-11/2 microcomputer for storage and subsequent processing.

3. The Stark Structure of Inhomogeneously-Broadened Spectra. Splitting Dispersion. Correlation of the Various Electron Transition Energies

The physical basis for using the selective laser excitation technique is the interrelationship between the properties of the electron transitions in the atoms (ions) and the composition, structure, and stresses of their nearest coordination shells (polyhedrons). In this respect all solids can be divided into two classes. The first class includes structurally-degenerate crystalline substances. Here either optically active ions form a perfect lattice, or the impurity or intrinsic defects form several types (once again, structurally-degenerate) optical centers. All centers are identical within one type. The second class includes structurally disordered matter. This class is characterized by a broad spectrum of spatial fluctuations in defect structure. This class includes disordered crystalline solid solutions and glasses. Their electron spectra are characterized by wide, inhomogeneously-broadened bands reflecting the integral optical properties of the activated medium.

The selective laser excitation technique makes it possible to successfully analyze solids belonging to both of these classes. However, if the former spectroscopic analysis techniques (such as the concentration series method [66]) were available for the first class of matter, traditional spectroscopy has yielded little for the second class. Until recently it appeared nearly impossible to recover any information on the properties of isolated defect groups, their interrelationships and interactions with the crystalline or glass-like base from the sum inhomogeneously-broadened spectra. Selective laser excitation, as well as special luminescence recording schemes and techniques coupled with special procedures for processing these data all make it possible to obtain extensive information on the properties of the medium and the microprocesses occurring under optical excitation.

3.1. Selective Spectroscopy of Inhomogeneously-Broadened Luminescence Bands of Rare-Earth Ions

The Eu^{3+} *ion.* Narrowband laser radiation acting on a structurally-disordered impurity material selectively excites isolated groups of optical centers and causes narrowing of the inhomogeneously-broadened spectral components and generates a Stark structure in the luminescence spectrum

We consider the spectroscopic properties of the Eu^{3+} ion which is most commonly used as a spectroscopic probe of the structure of optical material given its simple level scheme [67-69]. A comparison of the integral inhomogeneously-broadened luminescence spectra of the Eu^{3+} ions (the $^5D_0-^7F_0$, $^7F_1$ transitions) in three disordered bases ($ZrO_2-Eu_2O_3$, La-Al-silicate and Na-Y-B-silicate glasses with Eu^{3+}) has demonstrated that they are similar in shape. The only difference lies in the intensities and the average splitting scale. Monochromatic laser excitation of luminescence has been successful in revealing substantial differences in the structure of their inhomogeneously-broadened spectra reflecting differences in the formation of the spectral structure of the optical centers in these materials [57, 62]. Spectral and temporal selection techniques have revealed a discrete set (between 1 and 5 depending on the europium concentration) of optical centers with sharply differing spectral properties in $ZrO_2-Eu_2O_3$ crystals. Activated glasses are characterized by smooth, continuous variation in the spectral properties from one optical center to another.

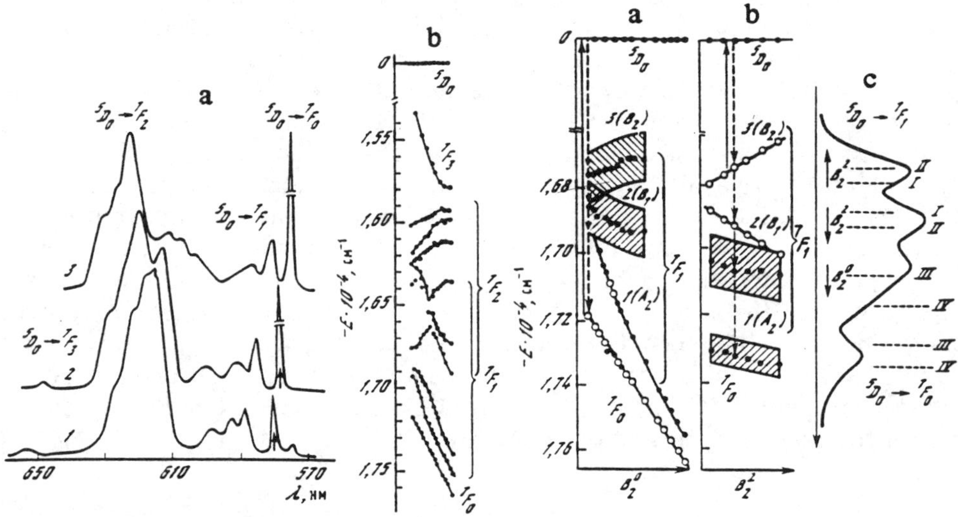

Fig. 3. Structure of inhomogeneous spectral broadening of Eu^{3+} in La-Al-silicate glass.

a - luminescence spectra under selective laser excitation (ν_h, cm^{-1}; 1 - 17,241, 2 - 17,301, 3 - 17,391); b - general diagram of Stark splitting of the levels $^7F_{0,1,2,3}$.

Fig. 4. Stark splitting diagrams of Eu^{3+} in Na-Y-B-silicate glass for Stokes (a) and anti-Stokes (b) excitation schemes and the inhomogeneously-broadened luminescence spectrum of Eu^{3+} (c).

Figure 3a shows the luminescence spectra of Eu^{3+} in La-Al-silicate glass (19 mass % La_2O_3, 27% Al_2O_3, 54% SiO_2) for various frequencies of the laser excitation radiation (indicated by the arrows) [57, 62]. Sharp details are clearly evident in the spectrum, while tuning the laser within the absorption transition reveals a substantial dispersion of the Stark components of these levels. A detailed picture of the variation in Stark splitting from center to center is shown in Figure 3b as a composite energy diagram. A differentiating feature of the spectral properties of the Eu^{3+} centers in La-Al-silicate glass was the correlated behavior of all energy levels analyzed: $^7F_0$, $^7F_1$, $^7F_2$, and $^7F_3$. Here the ligand field acting on the Eu^{3+} ions in this glass fluctuates over such a broad range that the Stark splittings of the $^7F_1$ and $^7F_2$ levels vary by an order of magnitude, reaching ~ 1000 cm^{-1} with a spin-orbital splitting of the ^7F state of 400-700 cm^{-1}.

A different picture is characteristic of Eu^{3+} ions in Na-Y-B-silicate glass (15 mass % Na_2O, 35% B_2O_3, 50% SiO_2) (Figure 4). We demonstrated in Refs. [50, 52, 56] that the structure of the inhomogeneously-broadened spectra in this matrix cannot be correctly represented by a single energy diagram. It was necessary to plot two different diagrams for the four Stark components of the $^7F_0$, $^7F_1$ (1, 2, 3) levels. Figure 4c shows an ordinary (nonselective) inhomogeneously-broadened luminescence spectrum for Eu^{3+} corresponding to them. We found a single group of spectra (centers) yielding the diagram shown in Figure 4a for the case of Stokes laser excitation of the $^5D_0$ level from the $^7F_0$ ground state or from the lower (I) Stark component of the $^7F_1$-level, while in the case of anti-Stokes excitation a different group of spectra (centers) and the diagram shown in Figure 4b (see centers I and II) were obtained from the upper two components (III, IV) of the $^7F_1$-level.

We note that we were able to expand the range of investigation of spectrally-different centers and to achieve selection at the far wings of the inhomogeneously-broadened $^7F_0-^5D_0$ band (optical density $d < 10^{-3}$) through frequencies exceeding the inhomogeneous distribution half-width by a factor of 6 [50, 52, 62] by using powerful pulsed excitation (a rhodamine 6Z laser) and a high-sensitivity photographic recorder. This same feature of our technique made possible an anti-Stokes scheme for exciting Eu^{3+} ions from the $^7F_1$ level (3) whose population is only 2% of the total even at $T = 300$ K.

The need to draft two different level diagrams, which was neglected in the spectral structure analysis by the authors of Refs. [36-39, 70-73], derives from the fact that excitation at the $^7F_0-^5D_0$ transition (center III or IV) serves to narrow the $^5D_0-^7F_1(1)$ band at the same time that the $^5D_0-^7F_1(2, 3)$ bands remain unchanged (they do not "sense" selection). Residual inhomogeneous broadening of the latter is due to the uncertainty in the position of $^7F_1(2)$ and $^7F_1(3)$ levels which is represented in Figure 4a by the hatched region. Anti-Stokes excitation at the $^7F_1 (3)-^5D_0$ transition (center I or II) narrows the $^5D_0-^7F_1(2)$ band analogously, although it does

14   *Alimov et al.*

not alter the $^5D_0-^7F_0$, $^5D_0-^7F_1(1)$ band. This suggests that the energy positions of the $^7F_0$ level (relative to $^5D_0$) and the lower Stark component $^7F_1(1)$ are correlated as are the energy positions of the two upper Stark components: $^7F_1(2)$ and $^7F_1(3)$.

Such complex behavior of the spectra can be described quantitatively if we introduce the electron transition correlation parameter a, which reflects their degree of interdependence; we introduced this parameter previously in Refs. [57, 62]. We define the correlation parameter as the ratio of the narrowing coefficients of the spectra at nonresonance (c_n) and resonance (c_r) (with respect to the monochromatic excitation radiation) transitions: $a = (c_n - 1)/(c_r - 1)$. Here the narrowing coefficient c represents to what degree the inhomogeneous luminescence spectral component has narrowed in the transition from ordinary excitation to laser excitation. Such an approach allows a quantitative determination of the degree of correlation of transition. The correlation parameter adopts values from $a = 0$ with a complete absence of correlation to $a = 1$ with complete correlation of the transitions.

The correlation parameters for three transitions of Eu^{3+} from the $^5D_0$ level to the $^7F_1(1, 2, 3)$ level in various glasses calculated relative to the $^5D_0-^7F_0$ resonance (excitation) transition are given below:

Transition	$^5D_0-^7F_1(1)$	$^5D_0-^1F_1(2)$	$^5D_0-^7F_1(3)$
a(La-Al-silicate)	0.34	0.15	0.15
a(Na-B-silicate)	0.65	0.05	0.05

It is clear that the La-Al-silicate glass is characterized by moderately significant, yet similar values of the parameters a for all three transitions, which permitted description of the properties of the Eu^{3+} ion using a single Stark splitting diagram. The differences in the degree of correlation of the electron transitions for the Na-Y-B-silicate glass exceeded an order of magnitude which made it necessary to draft at least two composite energy diagrams for the Eu^{3+} ions.

It is clear from these data that the standard Stark splitting scheme of Eu^{3+} ions which is a simple scheme for structurally-degenerate crystals [67, 69] is substantially more complex in crystalline solid solutions and glasses and is determined by their composition. However, this increasing degree of complexity may yield additional information from a spectroscopic analysis. Indeed, we cannot assign any irreducible representations (crystalline quantum numbers) to the energy levels in the crystals based on only a single Stark splitting diagram. Additional studies of line polarization or transition probabilities will be needed for this process [67-69]. The preliminary analysis may be simpler for glasses with a continuous set of centers.

For example, plots of the energies of components 1-3 of the $^7F_1$ level as a function of the frequency of the laser-selected centers shown in

Figure 4a, b within the framework of the rhombic symmetry of these centers permits unambiguous assignment of the representations B_2, B_1, and A_2 to the Stark components of the $^7F_1$-level. Indeed, the mutual dependence of the energies of the two upper Stark components of $^7F_1$(2 and 3) which does not correlate with the position of the lower component $^7F_1$(1) or the $^7F_0$ level, indicates that $^7F_1$(2, 3) originates from the doubly-degenerate level E in a tetra- or trigonal crystal field. Such a level identification makes it possible to attribute the independent variation of one of the two low-symmetry (axial or azimuthal) field strength parameters of the crystal field B_2^0 or B_2^2 to each of the diagrams in Figure 4a and b. Here the random variations of one of the parameters can be attributed to inhomogeneous spectral broadening of each transition.

The corresponding field strength parameter is indicated under each spectral component in Figure 4c, while the arrow indicates its direction of growth. The increasing energy of the shortwave component of the spectrum $^5D_0-^7F_1$(1) due to the increasing parameter B_2^0 can be attributed to the motion of one of the coordinating anion charges ($B_2^0 \sim R_{Eu-O}^{-3}$) along the principal axis of symmetry of the impurity complex. It is also possible to understand the correlation of the nephelauxetic shift (the centers of gravity of the $^7F_1$ and $^7F_0$ levels relative to $^5D_0$) attributable to the overlap of the wave functions of the rare earth ion-ligand and the parameter B_2^0 (see Figure 4a) within the framework of the smoothly varying coordination model of the Eu^{3+} ions in such glasses (from 8-9) [38, 39, 72] for the case where a ninth O^{2-} ion is introduced into the octuply-coordinated complex EuO_8. Thus, the diminishing distance between Eu^{3+} and the ninth ion which is responsible for the growth of the parameter B_2^0 may result in an expansion ("swelling") of the coordination polyhedron due to the repulsion of the O^{2-} ions, which thereby reduces the overlap of the wave functions of the Eu^{3+} ion with the remaining oxygen ions. At the same time, as we clearly see from Figure 4b, variations of the azimuthal parameter B_2^2 leave the mixing of the wave functions of Eu^{3+} and O^{2-} virtually unchanged.

Analogous concepts could also be applied to the spectral properties of the Eu^{3+} ions in La-Al-silicate glass. The significant difference in its structure derives from the fact that the Al^{3+} ions do not function as a modifier here which would result in a partial breakdown of the glass structure, but rather are introduced into its coordination grid with virtually no changes in rigidity [74]. The La^{3+} (Eu^{3+}) ions are often assigned the role of structural modifiers of the glass [74] although the literature also contains arguments in favor of the direct introduction into the glass-forming shell of the glass [75]. The stronger bond between the anion environment of the rare earth ion and the rigid shell of the La-Al-silicate glass is also manifested in the stronger interrelationship of different parameters of the crystal field as well as the greater Stark splitting and inhomogeneous broadening in this glass (see Figure 3).

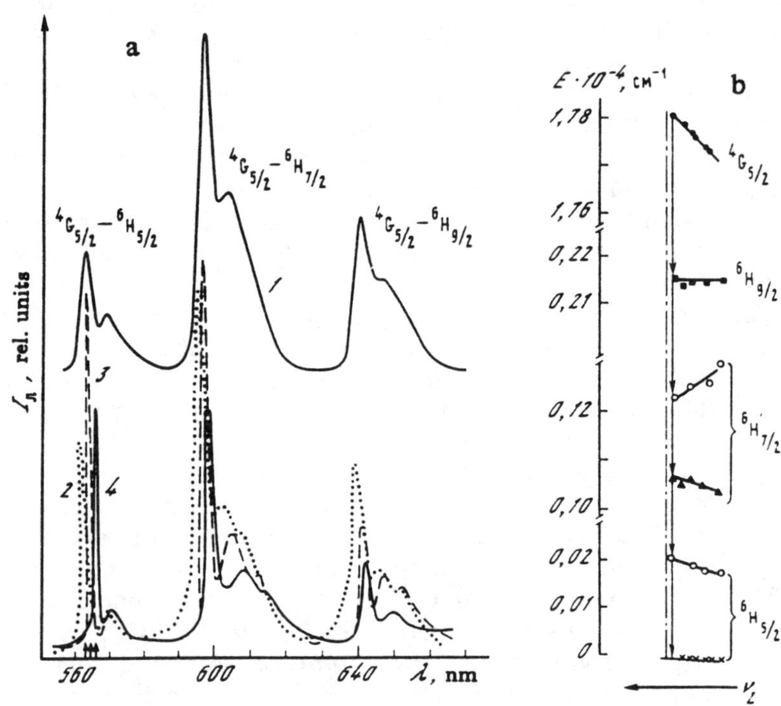

Fig. 5. Luminescence spectra (a) and composite diagram of Stark levels (b) of the Sm^{3+} ions.

a: 1 - nonselective excitation (incandescent lamp-filter), 2-4 - laser excitation (λ_L, nm; 2 - 562, 3 - 564, 4 - 566).

The Sm^{3+} *ion.* We used techniques analogous to those described above to investigate for the first time the selective luminescence spectra of Sm^{3+} ions in the matrix of La-Al-silicate glass [76]. The Sm^{3+} ion, like the Eu^{3+} ion, is a convenient model ion. The luminescent transitions in the visible portion of the spectrum and the fact that the metastable level lies within the band of the popular rhodamine 6Z laser makes Sm^{3+} ions useful as a spectroscopic probe of the structure of a crystalline or vitreous material and for investigating electron-phonon and ion-ion interactions in solids. Here the odd number of optical F-electrons of the Sm^{3+} ions which gives rise to half-integer values of the total quantum number J will yield a better (compared to the case of the Eu^{3+} ion) correlation of the spectral characteristics of the Sm^{3+} ions and the Nd^{3+}, Yb^{3+} ions: important activators for solid state laser media. A comparison of the spectral properties of these ions in similar matrices will help to define the energy structure of the most complex multiplets with a large number of heavily-overlapping states.

Figure 5a shows the selective and nonselective luminescent spectra of Sm^{3+} ions in La–Al–silicate glass. It is clear that selective laser excitation serves to narrow individual spectral components, causing substantial "structurization" of the inhomogeneously-broadened bands and reveals new inter-Stark transitions that are not detectable in the ordinary nonselective spectrum. This narrowing and the observed shifts of the spectral components of the bands (particularly the shortwave components) with changing frequency of the monochromatic excitation radiation (see arrows in Figure 5a) indicates a correlation of the energies of these transitions and makes it possible to plot an energy diagram for a number of Stark levels of the Sm^{3+} ion at different optical centers of the La–Al–silicate glass (see Figure 5b). This diagram reveals monotonic variations in the energies of the different levels of the optical centers selected at the $^6H_{5/2}$–$^4G_{5/2}$ transition for the case of monotonic tuning of the laser excitation radiation, which reflects a continuous variation of the field strength of the ligands in the transition from one Sm^{3+} center to another without the sudden jumps characteristic of variations in center symmetry [62, 77]. This fact combined with the results from a spectroscopic analysis of Eu^{3+} ions in La–Al–silicate glass outlined above suggest the absence in this matrix of microphases of substantially different symmetry which includes rare earth ions.

The Nd^{3+} *ion.* Fine structure studies of the spectra of Nd^{3+} ions in disordered laser matrices are of substantial interest and practical value. Inhomogeneous spectral broadening is responsible for the high efficiency and maximum energy output from such a laser as well as the spectral and temporal properties of the generated laser light. Figure 6 provides results from a spectroscopic analysis of Nd^{3+} ions for the case of selective laser excitation for two types of structurally-disordered matrices: LGS-28 silicate glass [54, 56, 78] and a crystalline solid solution of yttrium-fluorite CaF_2–YF_3 [77]. Using nonresonance selective excitation of Nd^{3+} ions in these media made it possible for us to identify for the first time a Stark structure of their inhomogeneously-broadened spectra and to draft splitting diagrams for the different groups of optical centers. The yttrium ions are chaotically distributed among the sites of the calcium sublattice while the excess fluorine ions occupy some of the octahedral interstitials in the case of a disordered CaF_2–YF_3 solid solution. As suggested from these data in this case the energy levels are classified in two subsystems (I and II) which indicates the formation in this solid solution of two subsystems of preferred coordination. Each of these experiences certain stresses (observed as inhomogeneous broadening), although there is no smooth transformation of one subsystem into another.

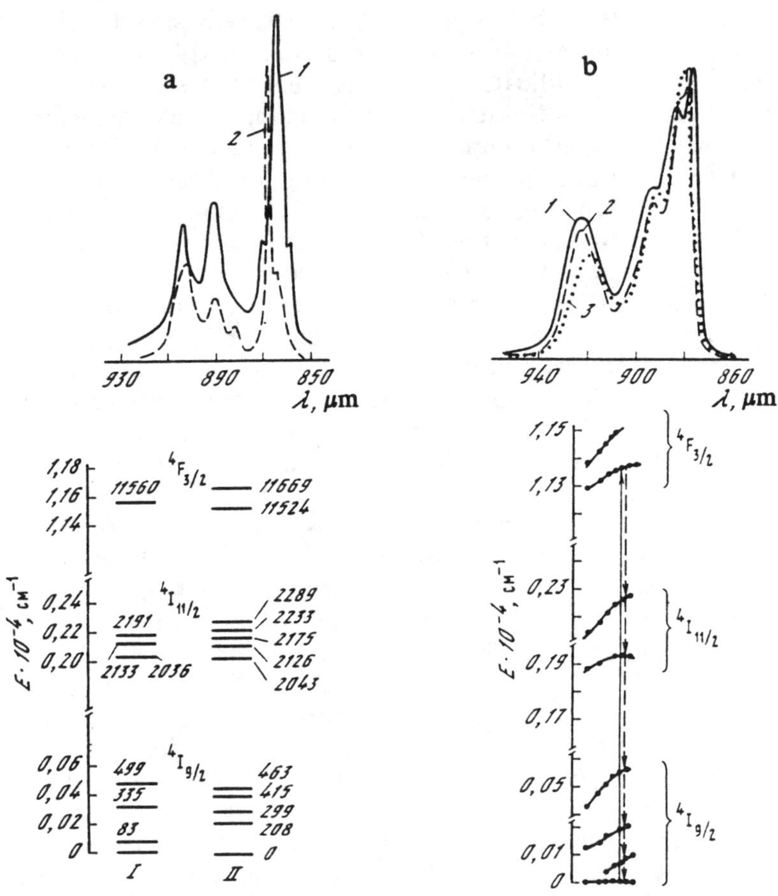

Fig. 6. Sections of the selective luminescence spectra of Nd^{3+} ions at the $^4F_{3/2}-^4I_{9/2}$ transition and the Stark splitting diagrams of the Nd^{3+} optical centers.

a - in a CaF_2-YF_3 crystalline solid solution (10%) (λ_L, nm: 1 - 579, 2 - 580); b - in LGS-28 silicate glass (λ_L, nm: 1 - 568, 2 - 570, 3 - 575).

A different situation is characteristic of activated laser glasses with Nd^{3+}. Here as in the case of the Eu^{3+} and Sm^{3+} ions in glass the transition from one optical center to another is smooth and continuous, although the nearest neighbor stresses of the impurity ions may differ quite substantially [57, 62].

It is interesting to compare the inhomogeneous broadening and Stark structure of Nd^{3+} spectra for the two primary classes of laser glasses: silicate and phosphate glasses. Figure 7 shows composite Stark splitting diagrams of the metastable $(^4F_{3/2})$ and ground $(^4I_{9/2})$ levels of Nd^{3+} ions in Li-La-phosphate [62, 79, 80] and LGS-28 silicate [78, 81, 82] glasses obtained by the selective laser excitation technique.[2] They are plotted relative to the center of gravity of the $^4F_{3/2}$-level in order to simplify the analysis.

It is clear from the diagram in Figure 7 that the LGS-28 glass has stronger variations (dispersion) in the energy of the lower level $^4I_{9/2}(0)$ compared to both components of the $^4F_{3/2}$-level (1 and 2). The dispersion range is given by the hatched and open rectangles to the right of the diagrams. The substantial energy dispersion of the lower level E_0 is manifested as inhomogeneous spectral broadening of both transitions (0-1 and 0-2) and is responsible for their correlation, although the magnitude of such broadening differs substantially. Indeed, as we see from the diagrams in Figure 7c the energy of the 0-2 transition is equal to

$$E_0 - E_2 = E_0^{c.g.} + E_0^{cr}(B_n^m) + E_2^{cr}(B_2^2, B_2^0),$$

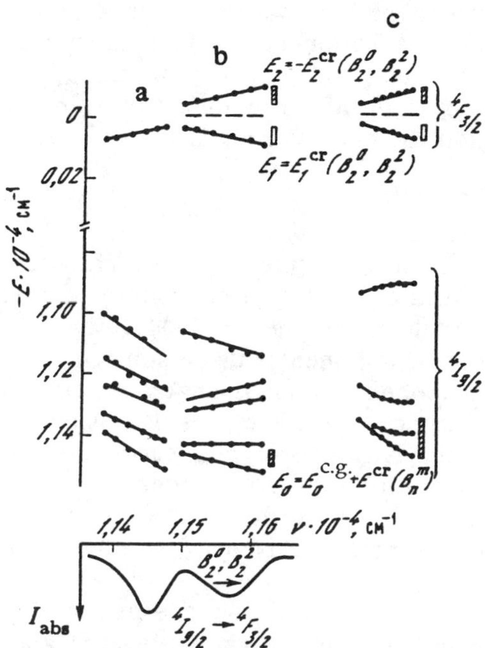

Fig. 7. Stark splitting diagrams of the $^4F_{3/2}$ and $^4I_{9/2}$ levels of Nd^{3+} in Li-La-phosphate glass (a, b) for the case of excitation at the 0-1 (a) and 0-2 (b) transitions and in LGS-28 silicate glass (c).

2. Because of the negligible inhomogeneous broadening of Nd^{3+} transitions in phosphate glass frequency selection of individual groups of Nd^{3+} optical centers was possible only by resonant excitation directly to the metastable $^4F_{3/2}$-level using the tunable LiF color center $(F_2 \rightarrow F_2^+)$ crystal laser discussed here [79, 83, 84].

i.e., the energy dispersion which determines the inhomogeneous broadening of this transition is obtained by geometric addition of the lengths of the hatched rectangles. Here $E_0^{\text{c.g.}}$ is the center of gravity energy of the $^4I_{9/2}$-level, while E_i^{cr} is the Stark splitting energy determined by the parameters of the crystal field B_n^m. For the 0–1 transition the energy is equal to

$$E_0 - E_1 = E_0^{\text{c.g.}} + E_0^{\text{cr}}(B_n^m) + E_1^{\text{cr}}(B_2^2, B_2^0),$$

while its dispersion (inhomogeneous broadening) is given by the difference of the Stark splitting energies (it is necessary to subtract the length of the open rectangle geometrically from the length of the shaded rectangle). Hence it is clear that the long-wavelength component will be substantially narrower than the short-wavelength component which is characteristic of the majority of glasses containing Nd^{3+}. Evidently this mechanism is common to inhomogeneous broadening for the majority of rare earth ions in glasses and can be traced by detecting the minimum inhomogeneous width for the longest wavelength components of the absorption bands and the shortest wavelength components of the luminescence bands (see Figures 5 and 6) for $T \to 0$.

In phosphate glass the variation in Stark splitting energy on the ground level (0) is of the same magnitude as on the metastable level (1 and 2). We can assume that these variations are due to the dispersion of the low-level symmetry parameters of the crystal field B_2^0, B_2^2 which cause splitting of the $^4F_{3/2}$ level. As in silicate and phosphate glass inhomogeneous broadening of the 0–2 transition (the dispersions are summed) is greater than for the 0–1 transition (the dispersions are subtracted). This is clear in the absorption spectrum shown in the lower section of Figure 7a, b. However, unlike silicate glass phosphate glass has different broadening of these two transitions. The energy of the 0–2 transition, like silicate glass, is determined by the sum of the low-symmetry crystal field parameters B_2^0 and B_2^2 on levels 0 and 2. The low-symmetry parameters are subtracted for the 0–1 transition and their identical variations from center to center on the metastable (1) and ground (0) levels compensate one another. This would also be manifested as a vanishing of the inhomogeneous broadening of the 0–1 transition if no other factors producing inhomogeneous broadening exist. Such factors include an independent change in the high-symmetry crystal field parameters B_4^m, B_6^m which affect the change in Stark splitting of the ground state $^4I_{9/2}$ which has no effect on the metastable level. Such an approach provides a plausible explanation for the low degree of correlation of the 0–2 and 0–1 transitions and the observed differences in the energy level diagrams (see Figure 7a, b) for the case of laser excitation to the lower (0–1 transition) and the upper (0–2 transition) Stark components of the metastable level $^4F_{3/2}$, respectively.

3.2. Selective Spectroscopy of the Inhomogeneously-Broadened Absorption Bands of Rare-Earth Ions

So far we have discussed studies of the internal structure of inhomogeneously-broadened luminescence spectra, electron levels below the radiative state of the rare-earth ions. However, from the viewpoint of a detailed spectroscopic analysis of the structure of an activated solid and in order to properly account for the spectral properties of the laser material it is necessary to know the Stark structure of levels above the metastable level. We therefore employed a number of laser-spectroscopic experimental setups [52, 57, 81, 82] that made it possible to reveal the Stark structure of the inhomogeneously-broadened absorption spectra. These setups in the stationary variant have been discussed and implemented in Refs. [18, 85, 86] devoted to the selective spectroscopy of organic molecules.

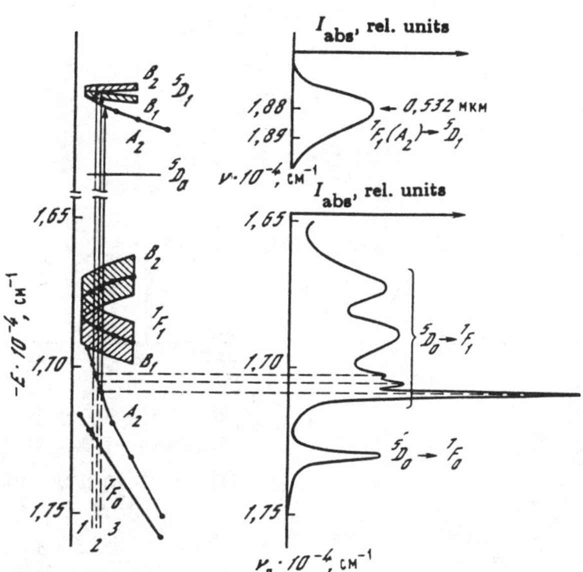

Fig. 8. The fine structure of the luminescence spectrum of Eu^{3+} ions in Na-Y-B-silicate glass for the case of nonresonant laser excitation and Stark splitting of the upper-lying "nonradiative" $^5D_1$-level.

The Eu^{3+} *ion.* Monochromatic second-harmonic radiation from an YAG-laser at $\lambda = 0.532$ μm in the absorption band of the $^7F_1(A_2)$—$^5D_1$ transition which has no structure in the integral spectrum (the absorption spectrum is shown in Figure 8 at the top right) was used in the first, simplest, experimental setup to obtain information on the Stark structure of the

$^5D_1$-level of Eu^{3+}. The lower right of this figure shows the nonstationary luminescence spectrum of the $^5D_0-^7F_1$ transition for Na-Y-B-silicate glass activated by Eu^{3+} (12 mass %) obtained by the excitation method described above for $T = 77$ K and with a small delay t_d to eliminate excitation migration ($t_d = 10^{-4}$ s) [52]. A sudden narrowing and a more complex short wavelength component of $^5D_0-^7F_1(A_2)$ (a triplet instead of a singlet) compared to the nonselective integral luminescence spectrum are also observed. As we see from Figure 8 (left) the recorded fine structure reflects a simultaneous selective laser excitation of three groups of optical centers (1, 2, and 3) essentially representing a fingerprint of the nonradiative $^5D_1$-level at the $^5D_0-^7F_1(A_2)$ luminescent transition. The scale of the fine (triplet) structure in the spectrum of $^5D_0-^7F_1(A_2)$ corresponds in a first approximation to the splitting scale of the $^5D_1$-level.

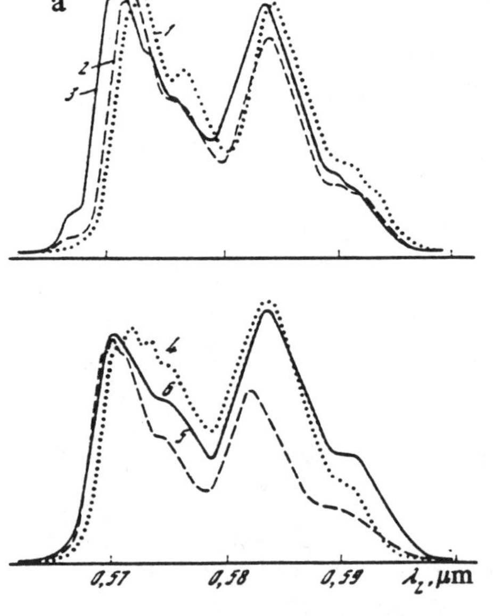
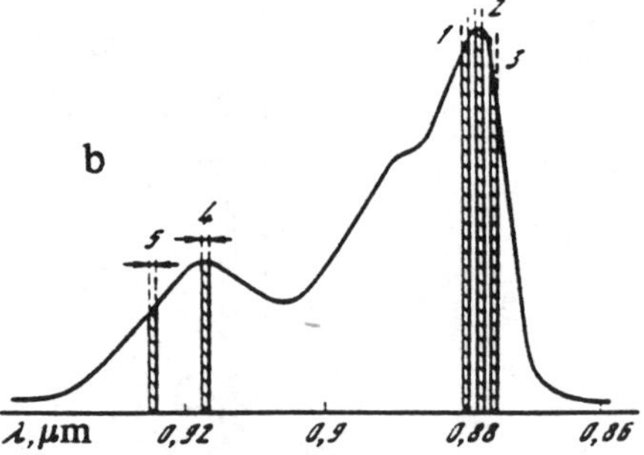

Fig. 9. Monochromatic excitation spectra of Nd^{3+} (curves 1–5) and the inhomogeneously-broadened absorption spectrum (curve 1) at the $^4I_{9/2}-^2G_{7/2}$, $^4G_{5/2}$ transition (a) for selective recording of luminescence at the short wavelength (1, 2, 3) and long wavelength (4, 5) spectral components of $^4F_{3/2}-^4I_{9/2}$ (b).

A comparison of the splitting scales of the two levels ($^7F_1$ and $^5D_1$) with an identical $J = 1$ makes it possible to find the ratio of equivalent crystal field operators for them. This ratio was 6.0, which substantially exceeds the theoretically value of 3.73 [67]. This can be understood as a violation of the weak crystal field condition for Eu^{3+} ions in the test glass (violation of the strict LS-coupling approximation; ordinarily calculations are carried out within the framework of this approximation [67]). The most sudden spectral narrowing of $^5D_0-^7F_1(A_2)$ at center 1 compared to centers 2 and 3 is easily explained within the framework of the lowest homogeneous broadening magnitude and the high degree of correlation of the $^5D_1(A_2) - ^5D_0$, $^5D_0 - ^7F_1(A_2)$ transitions at the same time that excitation was to the $^5D_1(B_1)$ and $^5D_1(B_2)$ levels for the remaining centers; the energy position of these levels relative to the $^5D_0$ level weakly correlates with the $^5D_0-^7F_1(A_2)$ luminescent transition.

Such a scheme using tunable laser excitation of the $^5D_1$ level was subsequently employed by the authors of Ref. [87] to develop a more comprehensive Stark splitting diagram of the $^5D_1$-level of Eu^{3+} (in the ion glass) and this demonstrated a good agreement with our results.

The Nd^{3+} *ion.* We employed selective laser excitation to analyze the structure of the upper-lying nonradiative $^2G_{7/2}$ and $^4G_{5/2}$ levels of the Nd^{3+} ions in silicate glass [81, 82]. Figure 9b shows the integral luminescence spectra of the $^4F_{3/2}-^4I_{9/2}$ transitions recorded under conditions of nonselective excitation. The hatched regions in these spectra represent a number of centers sequentially isolated by the monochromator (selective recording). The dependence of luminescence intensity on the wavelength of the monochromatic laser excitation was recorded for each of these centers (Figure 9a). The laser excitation frequency was continuously scanned in the wavelength range of the $^4I_{9/2}-^4G_{5/2}, ^2G_{7/2}$ absorption transition (the inhomogeneously-broadened absorption spectrum is also represented in Figure 9a (curve 6)). The complex structure of the Stark splittings of the upper-lying $^4G_{5/2}, ^2G_{7/2}$ states of the various optical centers of the Nd^{3+} ion was identified by the position of the excitation efficiency maxima $I(\lambda_L)$. These are shown in Figure 10. The stepped variation in the luminescence recording frequency v_r alters the monochromatic excitation spectrum, which reflects a shift in the position of the Stark sublevels, indicating a variation in the sublevels from center to center (see Figures 9 and 10).

Figure 10 presents the results of frequency selection of the luminescing centers at five Stark transitions between the $^4F_{3/2}-^4I_{9/2}, ^4I_{11/2}$ levels. The dashed lines identify transitions from the lower and upper Stark components of the $^4F_{3/2}$ state ($T = 77$ K) which correspond to the closed and open circles in the Stark splitting diagram, respectively. Energy coincidence of the $^4I_{9/2}-^4G_{52}, ^2G_{7/2}$ absorption transition may serve as a criterion for identifying the same group of centers at different luminescent transitions (see horizontal dashed lines 1-3 labeling the three groups of optical centers). By measuring the energy between the Stark levels of the same center on the X-axis it is possible to determine the energy gap $E(^4I_{11/2}-^4I_{9/2})$ for it together with the Stark splittings $\Delta E(^4I_{9/2})$ and

$\Delta E(^4F_{3/2})$. Therefore it is possible to determine for center 1 from Figure 10 the spin-orbital splitting energy $E(^4I_{11/2}-^4I_{9/2})$ together with the Stark splitting $\Delta E(^4F_{3/2})$, while it is possible to determine the Stark splittings $\Delta E(^4F_{3/2})$ and $\Delta E(^4I_{9/2})$ for center 2, although it is only possible to determine $\Delta E(^4I_{9/2})$ for center 3.

The analysis of the Stark structure of the fundamental luminescence and absorption inhomogeneously-broadened spectra described above made it possible to formulate a composite diagram of the most important levels of Nd^{3+} in the LGS-28 laser silicate glass [57, 62].

Figure 11 demonstrates the narrowing of the luminescence spectrum (curve 1) in the case of selective excitation of one group of centers with λ_L = 569 nm from the integral absorption spectrum (curve 2) relative to the integral luminescence spectrum (curve 3). Figure 11 also shows narrowing of the excitation spectrum (curve 4) with selective recording of luminescence of one group of centers λ_r = 878.6 nm from the integral luminescence spectrum (curve 3).

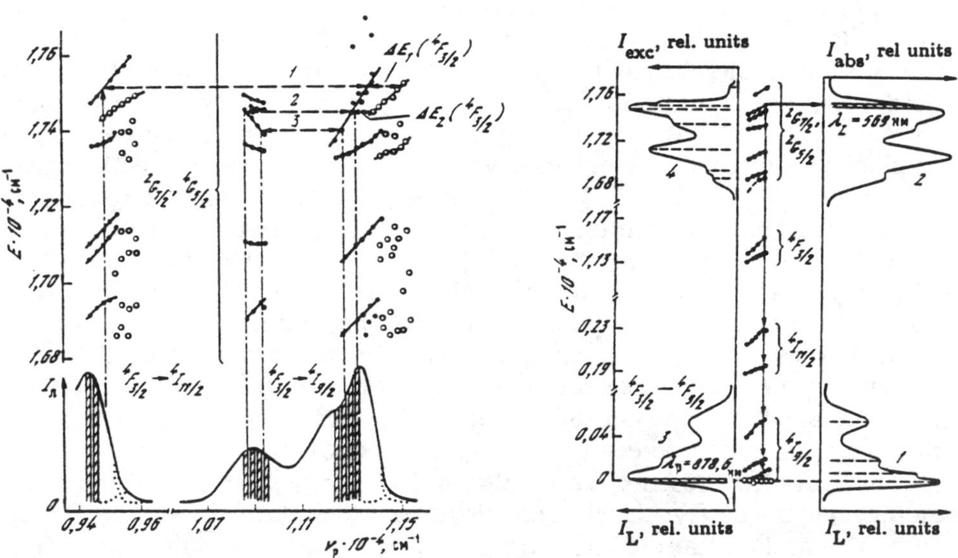

Fig. 10. Start splitting diagrams of the upper-lying nonradiative $^2G_{7/1}$ and $^4G_{5/2}$ levels of the Nd^{3+} ion in LGS-28 glass and their relation to the fundamental luminescent $^4F_{3/2}-^4I_{9/2}$, $^4I_{11/2}$ transition.

Fig. 11. Composite Stark splitting diagram of the $^4I_{9/2}$, $^4I_{11/2}$, $^4F_{3/2}$, $^4G_{5/2}$, $^2G_{7/2}$ levels for a continuous set of optical centers in LGS-28 laser silicate glass.

The central section in Figure 11 yields a composite picture of variations in Stark splitting from center to center (Nd^{3+}) for the $^2G_{7/2}$, $^4G_{5/2}$, $^4F_{3/2}$, $^4I_{9/2}$, $^4I_{11/2}$ levels which determine the most critical absorption, luminescent and lasing transitions of the Nd^{3+} ions in laser glass.[3] More complete knowledge of the fine structure of these transitions will make it possible to carry out more accurate calculations of quantum oscillators and amplifiers using such a structurally- and spectrally-disordered active medium.

The Yb^{3+} *ion.* The laser spectroscopy techniques developed and described above for application to the spectroscopy of inhomogeneously-broadened luminescent and absorption transitions were also employed in our case to investigate the spectral properties of Yb^{3+} in Ba-Al-phosphate glass (Figure 12). The Yb^{3+} ion has the simplest energy scheme consisting of only two terms ($^2F_{5/2}$ and $^2F_{7/2}$) linked by rather strong stimulated electric dipole transitions. This ion has therefore been a constant source of interest to spectroscopists and is used in formulating a number of important model experiments to investigate the optical manifestations of interionic and electron-phonon interactions in solids [4, 33, 94-97]. Moreover the Yb^{3+} ion is itself a rather efficient laser ion [2-4] and is also widely used as an effective sensitizer in a wide range of laser hosts [1-4] and in quantum infrared-to-visible light converters [98].

The actual possibility for a detailed structural analysis of inhomogeneous broadening of Yb^{3+} in disordered matrices arose in 1979 after the development of effective, reliable tunable LiF crystal lasers ($F_2 \rightarrow F_2^+$) [79, 83, 84] operating in the pulsed-periodic mode, producing a broad tuning range (0.84-1.1 μm) and a narrow lasing spectrum (0.65-2.0 Å).

Figure 12a provides our sample luminescence spectra of $^2F_{5/2}(1)$—$^2F_{7/2}(1, 2, 3, 4)$ obtained by selective, direct laser excitation of Yb^{3+} to the lower metastable level $^2F_{7/2}(1)$—$^2F_{5/2}(1)$ (the excitation frequencies ν_L are given by arrows 1-3). Figure 12b demonstrates the laser excitation spectra for the case of selective luminescence recording at the $^2F_{5/2}(1)$—$^2F_{7/2}(1)$ transition. As we see there are significant shifts of the Stark components in the luminescence and excitation spectra with varying excitation and recording frequency, indicating the existence of an energy correlation of the $^2F_{5/2}(1)$—$^2F_{7/2}(1, 2, 3, 4)$ and $^2F_{7/2}(1)$—$^2F_{5/2}(1, 2, 3)$ transitions. An analysis of a substantial number of similar selective spectra allowed us to draft for

---

[3] We note that to date the literature has not contained any data (other than ours) on the Stark splitting structure of the inhomogeneously-broadened absorption and luminescent bands of Nd^{3+} ions in glasses, although a number of studies have been devoted to their spectroscopy using selective excitation and recording techniques [88-93].

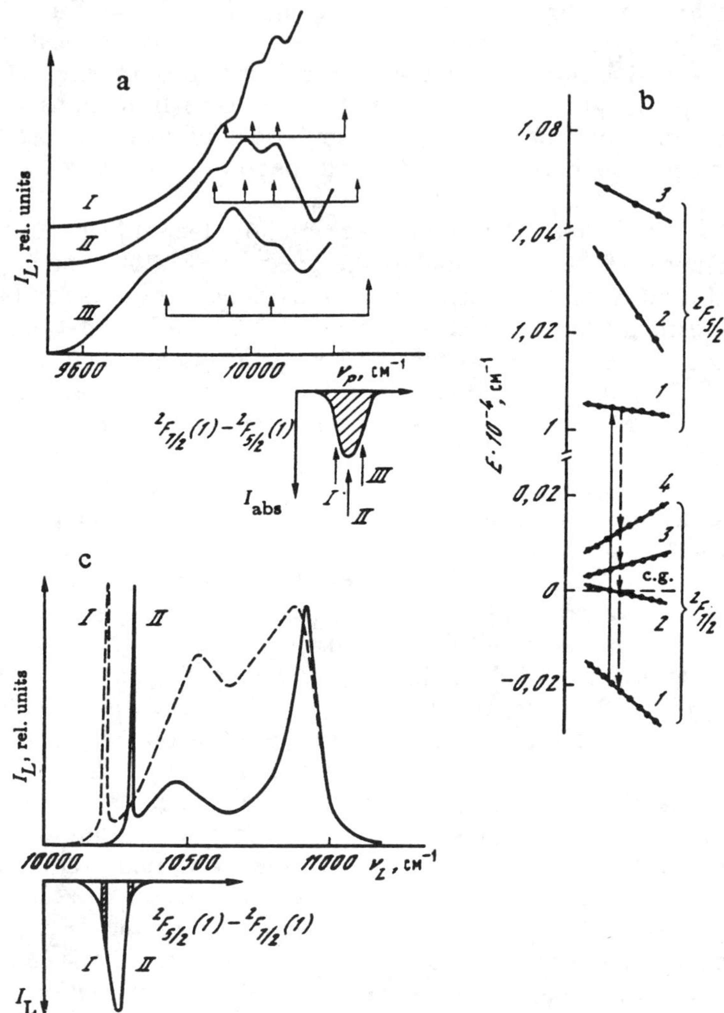

Fig. 12. Low-temperature (T = 77 K) luminescence spectra of Yb^{3+} at the $^2F_{5/2}$—$^2F_{7/2}$(1, 2, 3, 4) transitions under selective excitation of $^2F_{7/2}$(1) —$^2F_{5/2}$ (a) and the monochromatic excitation spectra of $^2F_{7/2}$(1)—$^2F_{5/2}$(1, 2, 3) in conditions of selective luminescence recording at the $^2F_{5/2}$(1)—$^2F_{7/2}$(1) transition (b) and a composite Stark splitting diagram of the $^2F_{5/2}$ and $^2F_{7/2}$ levels for a continuous set of Yb^{3+} optical centers in Ba–Al–phosphate glass (c)

a: I - v_L = 10,204 cm^{-1}, II = 10,246, III - 10,277; b: I = v_r = 10,203 cm^{-1}, II = 10,288 cm^{-1}

the first time a comprehensive composite Stark splitting diagram[4] (relative to the center of gravity of the $^2F_{7/2}$ term) indicating its dispersion for a continuous set of Yb^{3+} optical centers in Ba-Al-phosphate glass (Figure 12c).

4. Structure of the Vibronic Transitions in Inhomogeneous Spectral Broadening

Fine structure (Stark) studies of the inhomogeneously-broadened bands of electron transitions (the zero-phonon spectral lines) are, from the methodological viewpoint, closely related to investigations of inhomogeneously-broadened vibronic transitions (the phonon wings). An analysis of the shape of the phonon wings of both the absorption and luminescence spectra of impurities in solids represents an important optical technique for investigating the vibrational states of the matrix [100-109]. A purely electronic (zero-phonon) transition for condensed matter with a high degree of structural order may be fixed for impurity centers accurate to better than 0.1-1 cm^{-1} (minor inhomogeneous spectral broadening of the transition). This makes possible a highly precise determination of the energies of phonons involved in the vibronic transitions by counting from the zero-phonon line. The differential between the local crystalline fields of individual impurity centers for solids with a disordered structure (compound crystals, glasses, and liquids) causes a substantial dispersion of the electron transition energy which is manifested as a significant inhomogeneous broadening of their spectra, which reaches several hundred cm^{-1}. An accumulation of a large number of vibronic spectra relatively shifted along the energy scale yields little information on the collective inhomogeneously-broadened vibronic spectrum. Therefore, as in the case of the electron states, in order to recover the desired information on the vibrational spectrum it is possible to use selective laser luminescence excitation and spectroscopy of monochromatic excitation by selective light recording.

The theoretical principles of applying such selective techniques to the vibronic transitions have been rather well developed [18, 110-112] while the number of experimental observations for complex molecules has been substantial (see Refs. [17, 18] and the references cited therein). However, it has turned out to be difficult to recover the true shape of the phonon wing based on observed spectra for systems with strong electron-phonon coupling

---

[4] We believe that the only study where such a problem for a different (silicate) class of glasses containing Yb^{3+} ions has been numerically solved is Ref. [99] which employed traditional (nonlaser) frequency selection techniques for application to the Yb^{3+} ions which have poor spectral and temporal resolution.

[18, 110-112]. The data from selective spectroscopy of the phonon wings of rare earth ions for which weak electron-phonon interaction occurs are not generally available.

Refs. [4, 113-115] have attempted to determine the vibrational state spectrum in glasses containing rare earth ions by measuring the effectiveness of nonresonant ion-ion interactions in which the host oscillations acquire the excess electron energy expended in the center-to-center transfer of excitation. This fundamentally interesting phenomenon is indisputably worthy of extensive analysis (see the second paper in the present volume), although in view of its complexity and relative neglect it requires comparison to more direct techniques.

We proposed a simple method of obtaining information on the vibrational spectrum of a disordered medium in Ref. [116]. This method is based on analyzing the type of inhomogeneously-broadened luminescence spectrum of the electron transition under excitation by fixed-frequency monochromatic light in the overlap range of the electron and vibronic (V) states of many centers. Methodologically such a scheme is analogous to that shown in Figure 8 with the difference that the vibronic states in this case function as the test electron $^5D_1$ level.

The inhomogeneously-broadened spectrum of the transition (for example, the absorption spectrum) is shown to the left in Figure 13a; this reflects the distribution of optical center number $n(v)$ as a function of their zero-phonon transition energy. We found laser excitation resulted in double selection of the optical centers: the first was resonant selection of centers with an electron energy equal to that of the excitation light (the narrow luminescence peak to the right in Figure 13a) while the second selection is nonresonant selection of the centers in the Stokes range from resonance excited by the vibronic transitions (the broad peak). Here it is obvious that the vibronic state density at the excitation frequency is determined by the product of the electron state density $n(v)$ and the vibrational state density $\rho(\epsilon)$. This appears as a dip in the luminescence spectrum near the low phonon frequencies where the phonon density vanishes at a high electron state density.

We demonstrate the possibility of such an analysis based on a sample investigation of the inhomogeneously-broadened spectrum of the electron transition between the lowest Stark components of the metastable $^2F_{5/2}(1)$ state and the ground $^2F_{7/2}(1)$ state of the Yb^{3+} ion in Ba-Al-phosphate glass.

The luminescence spectra were measured by using a tunable laser operating at the F_2^+ color centers in LiF as the monochromatic excitation source. The low concentration of Yb^{3+} ions in the specimen ($2.9 \cdot 10^{19}$ cm^{-3}) and temporal strobing of the luminescence signal ($t_d = 10^{-4}$ s) were selected to eliminate ion-ion interaction processes which cause energy transfer from one optical center to another and distort the spectra. The measurements were carried out at T = 4.2 K in order to eliminate the effect of homogeneous broadening and stimulated phonon emission processes on the derived spectra. For rare earth ions where the weak electron–phonon coupling approximation holds (the Debye–Waller factor α is near unity for the Yb^{3+} ions in the glass; α > 0.95) and neglecting the effect of the phonon wing on the luminescence spectrum it is possible to represent an arbitrary inhomogeneously-broadened luminescence contour as a product of the spectral functions:

$$I(v) = n^*(v)A(v)hv. \tag{1}$$

Here hv is the luminescence photon energy; $A(v)$ is the frequency dependence of the radiative transition probability (when it varies from center to center); $n^*(v)$ is the distribution of the excited centers over their electron transition frequencies. When an equal-probability technique is used to excite the different optical centers comprising the inhomogeneously-broadened band, the distribution $n_1^*(v)$ coincides with the frequency distribution of the centers in the ground state:

$$n_1^*(v) = kn(v), \tag{2}$$

while the luminescence contour is identical to the ordinary inhomogeneously-broadened absorption contour of this transition measured at a low optical density. Such a contour is represented by curve 1 in Figure 13b for Yb^{3+} ions in glass.

Another situation occurs in selective laser luminescence excitation of Yb^{3+} (see Figure 13b, curve 2). Here two sections are observed in the distribution function $n^*(v)$: a narrow spectral peak corresponding to the resonant excitation of the centers whose electron transitions are identical to the laser frequency ($v_{max} = v_L = 10,298$ cm^{-1}) and a broad distribution $n_1^*(v)$ in the Stokes range of the spectrum ($v_{max} = 10,235$ cm^{-1}) corresponding to nonresonant excitation of the Yb^{3+} optical centers by laser light through the vibronic absorption wing.[5] We note that in spite of the fact that the width of distribution $n_2^*(v, v_L)$ (curve 2) is similar to that of distribution $n_1^*(v)$ (curve 1) their shape is substantially different and this reflects the selectivity of laser excitation through the vibronic wing.

---

[5] The Stokes inhomogeneously-broadened spectrum resulting from excitation at the high-frequency wing of the absorption band from strong electron-phonon interaction has been discussed in Refs. [110, 117], while we observed the Stokes inhomogeneously-broadened spectrum in the case of weak electron-phonon coupling for Eu^{3+} ions in Na-Y-B-silicate glass, although this spectrum was not analyzed.

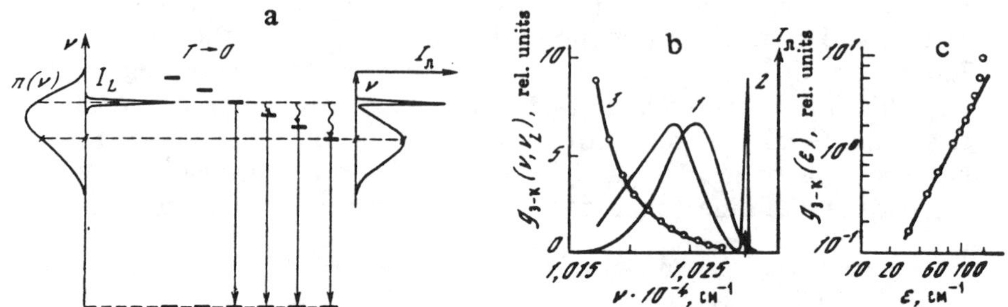

Fig. 13. Method of obtaining information on the vibronic spectrum of a disordered medium employing Yb^{3+}-activated Ba-Al-phosphate glass.

a - scheme for selective excitation and recording of the vibronic state density in the disordered medium based on the luminescence spectrum of the Yb^{3+} ions in Ba-Al-phosphate glass; b - the spectral relations: 1 - $I_1(v)$ (luminescence spectrum in the case of equal-probability excitation proportional to the distribution $n(v)$ of center number over their electron transition frequencies); 2 - $I_2(v, v_L)$ (luminescence spectrum under monochromatic excitation at v_L proportional to the center distribution $n(v)$ in the ground state and the vibronic spectrum $g_v(v, v_L)$ (curve 3)); c - the form of the vibronic spectrum $g_v(\epsilon)$.

It is clear from a comparison of curves 1 and 2 in Figure 13b that the centers whose transitions are close to the laser frequency v_L are excited with substantially lower efficiency than centers at remote frequencies. These differences are determined by the form of the function $n_2^*(v, v_L)$ which is expressed as the product of $n(v)$ and the desired vibronic spectrum:

$$n_2^*(v, v_L) \sim n(v)g_v(v, v_L). \tag{3}$$

In this expression $g_v(v, v_L)$ determines the absorption efficiency of laser radiation at frequency v_L at the vibronic transition of the optical center which has an electron transition at frequency v. A comparison of expressions (3) and (2) reveals that dividing spectrum I_2 by spectrum I_1 yields the desired form of $g_v(v, v_L)$ which remains undistorted by inhomogeneous broadening:

$$g_v(v, v_L) = I_2(v, v_L)/I_1(v). \tag{4}$$

In such a vibronic spectrum the vibrational frequencies of the matrix $\epsilon = v_1 - v$ can be determined with a high degree of accuracy since laser excitation fixes the frequency of the vibronic transition v_L accurate to the laser linewidth $\Delta v_L = 0.01 - 1$ cm^{-1}, while the continuously-varying electron transition frequencies v for the impurity ions in the glass are fixed accurate to the magnitude of homogeneous broadening which also is quite small at $T = 4.2$ K: $\gamma \ll 0.01$ cm^{-1}.

The form of the function $g_v(v, v_L)$ for Ba-Al-phosphate glass with Yb^{3+} calculated using Expression (4) is shown in Figure 13b (curve 3). The relation $g_v(\epsilon)$ corresponding to it is given on logarithmic coordinates in Figure 13c. The linear growth of this function with a slope of 2 indicates its exponential nature $g_v \sim \epsilon^2$ similar to the known spectrum of low-frequency acoustical phonons in the Debye model ($\rho \sim \epsilon^2$). It is interesting that by varying the laser excitation frequency relative to the maximum of the frequency distribution $n(v)$ we can substantially alter the vibrational frequency range analyzed by such a method: $\epsilon = v_L - v$.

We note that the vibronic spectrum $g_v(\epsilon)$ derived here will coincide with the distribution function of the vibrational frequencies of the matrix $\rho(\epsilon)$ only if the radiative electron transition probability and the electron-phonon coupling constant remain unchanged in the run from one optical center to another, while the latter is independent of the frequency of the phonon involved in the vibronic transition.

Media with impurity and intrinsic defects whose transitions correspond to strong electron-phonon coupling have been the subject of substantial interest in recent years in connection with the problem of fabricating tunable solid state lasers. In this case the widths of the vibronic absorption and luminescence spectra which often behave as quasihomogeneous spectra reach several thousand cm^{-1}. However, even such substantial "homogeneous" broadening often does not make it possible to neglect its inhomogeneous components. This is due to the substantially stronger dependence of the transition energy on the intracrystalline field strength (and its deformations) in such centers. The isolation and measurement of the homogeneous and quasihomogeneous broadening components of the vibronic transition represent a subject of substantial interest for predicting the laser properties of an active medium.

We carried out measurements of the luminescence spectra of Cr^{3+} ions in La-Al-borate glass [118, 119] and F_2^+-color centers in LiF under conditions of selective laser excitation to the vibronic absorption band.[6] Continuous shifts of the luminescence bands of 60 cm^{-1} for the Cr^{3+} ions in La-Al-borate glass and 50 cm^{-1} for the F_2^+-color centers in LiF were identified with increasing excitation frequency from 0.59 to 0.62 μm. This suggested a correlation between the absorption and luminescence vibronic

---

[6] Similar selective luminescence measurements of the following ions: Mo^{3+} [120], Cr^{3+} [121], Bi^{3+} [122], Ce^{3+} and Eu^{3+} [123] in glasses made it possible for the authors to identify a correlation in the spectral distribution of broad vibronic absorption bands and the luminescence of different optical centers in the glass.

transitions and made it possible to estimate the level of "homogeneous" and "inhomogeneous" broadening for F_2^+-CC in LiF crystals.

5. Luminescence of the F_2^+-Color Centers in LiF Crystals Under Selective Laser Excitation

The absorption and luminescence spectra of color centers (CC) in ionic crystals are broad amorphous bands for which the transitions between the vibrational sublevels of the ground and excited electron states are responsible. Normally the vibronic bands behave as inhomogeneously-broadened bands in lasing, thereby making it possible to completely eliminate the population inversion at a lasing spectrum that is several orders of magnitude narrower than the width of the luminescence spectrum.

Refs. [124-126] therefore report attaining monomode lasing at the F_A(II)-centers in KCl-Li and RbCl-Li and the F_B(II)-centers in KCl-Na and RbCl-Na crystals at $T = 77$ K at a power level amounting to 80% of the multimode lasing power.

At the same time we know that the relative position of the electron terms of the color centers and, therefore, their spectral luminescence properties are very sensitive to distortions in the local crystalline field of these centers. Such distortions occur when the crystal contains extraneous impurities or defects.

The spectral properties of the F_A(II)-centers therefore are substantially different from the F_A(I)-centers in that they have a substantially larger Stokes shift and narrower luminescence bands [127]. These differences arise from the substitution of the Na cation neighboring the anion vacancy with the Li cation. Strong variations in the position of the spectral lines of the F_2^+-centers and their half-widths have also been observed by the authors of Refs. [120, 129] from introduction of O_2, Mg, Ni, Ti, etc. impurities into LiF.

Therefore, the intrinsic or impurity crystal defects may cause a relative shift of the ground and excited levels of the color centers, thereby generating inhomogeneous spectral broadening. Such broadening may in turn substantially reduce laser efficiency in narrowband lasing conditions, particularly at low temperatures when the homogeneous component of the luminescence spectrum of the center is minimized.

We investigated the luminescence spectra of F_2^+-color centers in LiF crystals (the $2p\sigma_u \rightarrow 1s\sigma_g$ transition) under selective laser excitation from 0.532 to 0.62 μm. The measurements were carried out on two types of crystals: an LiF(F_2) crystal with stable neutral F_2^+-CC in which the F_2^+-centers were produced by two-stage photoionization of the F_2-centers at room temperature and LiF-O^2 (F_2^+), where the F_2^+-centers were stabilized by adding an oxygen impurity [128, 130]. Second harmonic radiation from a YAG:Nd^{3+} laser and radiation from a tunable rhodamine 6Z dye laser were employed to excite luminescence.

The luminescence spectra of F_2^+-color centers at $T = 77$ K are shown in Figure 14 (these spectra were corrected to account for the spectral sensitivity of the spectrometer). It is clear that a change in the excitation wavelength from 0.59 to 0.62 μm causes a near 50 cm^{-1} long-wavelength shift of the luminescence spectra of the F_2^+-centers (a change from 0.532 to 0.62 μm produces a shift of approximately 100 cm^{-1}), indicating inhomogeneous broadening of the luminescence transition.

We note that the nature of the shifts of the luminescence bands is similar in crystals with both unstable and oxygen-stabilized F_2^+-color centers. However, LiF-O^{2-}(F_2^+) crystals have a somewhat broader luminescence spectrum.

These results can be analyzed based on the configurational curves of the ground and excited states of the optical centers, where inhomogeneous spectral broadening is determined by the various positions of the excited electron levels relative to the ground level.

We examine the two limiting cases of configurational curves for optical centers that describe in simple terms the various manifestations of inhomogeneous broadening in the absorption and luminescence spectra of the color centers. The simplifying assumption is that the force constants of the adiabatic potentials of the excited electron states remain unchanged from center to center.

Figure 15a demonstrates inhomogeneous broadening of a group of spectrally diverse optical centers expressed solely as a variation in the configurational coordinates of the minima of the adiabatic potentials of the excited electron states of these centers with their energy position remaining unchanged. We will call this phenomenon "configurational" broadening. Figure 15b shows a "potential" inhomogeneous broadening model in which the various optical centers (from the group examined here) have identical configurational coordinates and differ only in the potential energy of the minima of the excited electron states. According to Figure 15a the "configurational" inhomogeneous broadening of the optical centers will be characterized by an anticorrelation of the absorption and luminescent transitions. This means that the lesser energetic (longer wavelength) centers in luminescence ($E_d'' < E_2'' < E_1''$) correspond to the more energetic (shorter wavelength) centers in absorption ($E_d' > E_2' > E_1'$). (Here and henceforth the primed quantities will represent absorption of the optical center while double primed quantities will refer to luminescence.) Such broadening is characterized by a dispersion of the Stokes shift, which varies from center to center. The "potential" broadening model (see Figure 15b), on the other hand, describes the correlated behavior of the absorption and luminescence transitions of the optical centers. Here the shorter wavelength centers in absorption have shorter wavelength luminescence.

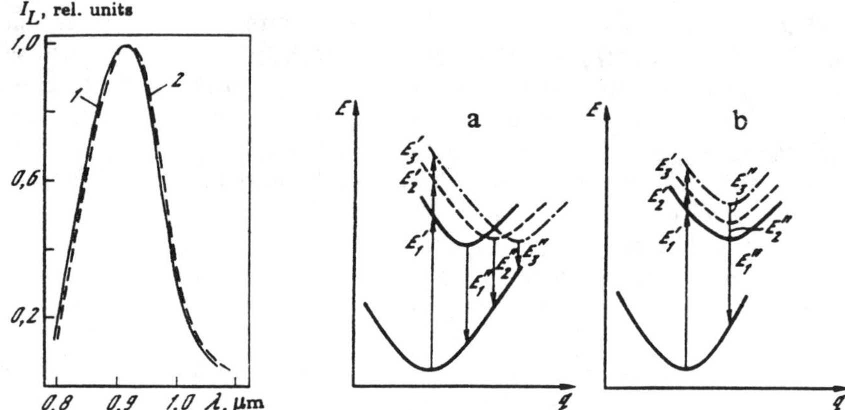

Fig. 14. Luminescence spectra of F_2^+-CC in LiF under monochromatic laser excitation

1 - $\lambda_L = 0.59$ μm; 2 - $\lambda_L = 0.62$ μm.

Fig. 15. Models of "configurational" (a) and "potential" (b) inhomogeneous broadening of optical centers.

As indicated by an analysis of the experimental data (see Figure 14) demonstrating a correlated shift in the luminescence spectra with excitation frequency the second broadening mechanism which is due to the change in potential energies of the transitions from center to center is the closest to the experimental situation occurring for F_2^+-color centers in LiF crystals.

We use the approach of the authors of Refs. [46, 110–112] to analyze the inhomogeneously-broadened spectra of the vibronic transitions. We investigate the relation between the shape and position of the absorption and luminescence spectra of a set of noninteracting centers (a medium of low optical density) with such internal characteristics of these spectra as homogeneous (δ) and inhomogeneous (Δ) broadening.

We examine an individual optical center with quasihomogeneous multiphonon vibronic absorption and luminescence bands with intensities I'_i and I'', respectively, which are satisfactorily described by Gaussian curves in the case of "intense heat liberation" [103]

$$I'_i (v' - v'_i) = \exp\left[-\frac{4\ln 2}{(\delta')^2} (v' - v'_i)^2 \right], \tag{5}$$

$$I_i'' (v'' - v_i'') = \exp\left[-\frac{4\ln 2}{(\delta'')^2} (v'' - v_i'')^2 \right], \tag{6}$$

where v_i', v_i'' are the maxima while δ', δ'' are the "homogeneous" widths of the absorption and luminescence bands, respectively, for the ith center.

It is possible to introduce inhomogeneous broadening of the absorption spectrum of the group of optical centers as a Gaussian statistical distribution of the centers over the frequencies of the maxima of their absorption transitions

$$g' (v_i') = \exp\left[-\frac{4\ln 2}{(\Delta')^2} (v_i' - v_{\max}')^2 \right] \tag{7}$$

where $v_{\max}'$ is the frequency corresponding to the maximum of the inhomogeneous distribution of the frequencies of the maxima v_i'; Δ' is the inhomogeneous width in the absorption spectrum

Summing the absorption signals from the various centers at frequency v' for the collective absorption spectrum we obtain

$$I_c' (v') = \int_0^\infty g' (v_i') I_i' (v' - v_i') \, dv_i'. \tag{8}$$

After substitution of Expressions (5) and (7) into Expression (8) this expression becomes the convolution of two Gaussian functions and, as demonstrated by integration, can also be Gaussian in form:

$$I_c' (v') = \text{const} \cdot \exp\left[-\frac{4\ln 2}{(\gamma_{\kappa}')^2} (v' - v_{\max}')^2 \right] \tag{9}$$

of width γ_c' related to the homogeneous width (δ') and the inhomogeneous width (Δ'), respectively, as

$$(\gamma_c')^2 = (\delta')^2 + (\Delta')^2. \tag{10}$$

In the same manner as (7) we introduce inhomogeneous broadening of the luminescence spectrum of the optical centers as a Gaussian statistical distribution of the centers over the frequencies of the maxima of the luminescent transitions

$$g'' (v_i'') = \exp\left[-\frac{4\ln 2}{(\Delta'')^2} (v_i'' - v_{\max}'')^2 \right] \tag{11}$$

where $v_{\max}''$ is the frequency of the maximum in the frequency distribution of the maxima v_i''.

Summing the luminescence signals of the various centers at v'' under conditions of equal-probability excitation yields expressions analogous to (8) and (9) for absorption:

$$I''_c (v'') = \int_0^\infty g'' (v''_i) I''_i (v'' - v''_i) dv''_i , \tag{12}$$

$$I''_c (v'') = \text{const} \cdot \exp \left[- \frac{4\ln 2}{(\gamma''_{\kappa})^2} (v'' - v''_{\max})^2 \right], \tag{13}$$

where

$$(\gamma'_c)^2 = (\delta'')^2 + (\Delta'')^2. \tag{14}$$

In the case of monochromatic excitation of frequency v_L the sum luminescence signal at v'' is given as

$$I''_M (v'') = \int_0^\infty [g' (v'_i) I_i (v'_L - v'_i)] I_i (v'' - v''_i) dv''_i . \tag{15}$$

We carry out integration of Expression 15 for three approximate "potential" inhomogeneous broadening models for the set of optical centers, determining the spectral width under monochromatic excitation γ''_m.

Model I. The force constants of the adiabatic potentials of the ground and excited states are identical for all optical centers ($\delta'_i = \delta''_i = \delta'_j = \delta''_j = \delta$) and $\Delta' = \Delta'' = \Delta$.

Model II. The force constants of the excited state differs from the force constant of the ground state although they remain unchanged from center to center ($\delta'_i = \delta'_j = \delta'$, $\delta''_i = \delta''_j = \delta''$, and $\Delta' = \Delta'' = \Delta$).

Model III. The force constants of the ground and excited states, as in model 1, are identical for all optical centers: $\delta'_i = \delta''_i = \delta'_j = \delta''_j = \delta$. The configurational coordinates of the adiabatic potential maxima vary from center to center, i.e., $\Delta' \neq \Delta''$, although "potential" inhomogeneous broadening predominates over "configurational" broadening. The inhomogeneous distributions of the frequency of the maxima at the absorption and luminescence transitions have a linear relation:

$$v''_i - v''_{\max} = k(v'_i + v'_{\max}).$$

Solving Equations (8), (12) and (15) for these simplified models we obtain relations that establish a relationship between the observed widths γ'_c, γ''_c and γ''_m and the characteristics of the elementary centers δ', δ'', Δ', and Δ'':

$$\text{for model I} \quad (\gamma^1_{\mathsf{T}})^2 = \delta^2 + \Delta^2, \quad (\gamma''_{\kappa})^2 = \delta^2 + \Delta^2,$$

$$(\gamma''_M)^2 = \frac{\delta^2(\delta^2 + 2\Delta^2)}{\delta^2 + \Delta^2} ;$$

for model II $\quad (\gamma_c')^2 = (\delta')^2 + \Delta^2, \quad (\gamma_{\kappa}'')^2 = (\delta'')^2 + \Delta^2,$

$$(\gamma_{M}'')^2 = \frac{2\,(\delta')^2\,(\delta'')^2 + \Delta^2\,(\delta'')^2 + \Delta^2(\delta')^2}{\delta'^2 + \Delta^2} \;;$$

for model III $\quad (\gamma_c')^2 = \delta^2 + (\Delta')^2, \quad (\gamma_{\kappa}'')^2 = \delta^2 + k^2\,(\Delta')^2,$

$$(\gamma_{M}'')^2 = \frac{\delta^2\,[\delta^2 + (\Delta')^2\,(1 + k^2)]}{\delta^2 + (\Delta')^2}$$

Substituting the experimental values of $\gamma_c' = 2700$ cm^{-1}, $\gamma_c'' = 1900$ cm^{-1}, $\gamma_m'' = 1800$ cm^{-1} into these solutions we find approximate values for the homogeneous and inhomogeneous broadening components of F_2^+-CC in LiF (Table 2). The actual behavior of the configurational curves of F_2^+-CC in LiF is unknown and may be different from the simplified models considered here. These values should therefore be considered a first estimate. However, it appears that this result provides a correct reflection of inhomogeneous broadening within an order of magnitude; in this case it is near the homogeneous broadening level at low temperatures.

Table 2

Calculation of homogeneous and inhomogeneous components of the absorption and luminescence spectra of F_2^+-color centers in LiF at $T = 77$ K.

Parameter	Model I	Model II	Model III
δ', cm^{-1}	1500	2400	1700
δ'', cm^{-1}	1500	1400	1700
Δ', cm^{-1}	1150	1300	2100
k	1	1	0.4
$\Delta'' = k\,\Delta'$, cm^{-1}	1150	1300	840
δ''/Δ''	1.3	1.1	2

The room temperature studies demonstrated that in this case the spectral width γ_m'' increases by a factor of approximately 1.3. In this case no shifts are detected in the luminescence spectrum from a changing excitation wavelength. This can be attributed to the increase in the homogeneous spectral component with increasing temperature (by a factor of approximately 1.45 with a rise in temperature from 77 to 300 K) due to the enhancement of electron-phonon interaction as well as the decrease in static inhomogeneous broadening attributable to the freezing of the various crystal centers. The luminescence band of F_2^+-color centers at 300 K is therefore more homogeneous than at 77 K and consequently room temperature can be assumed to be preferred for tunable lasing using LiF crystals containing F_2^+-color centers.

6. Homogeneous Broadening of Inhomogeneously-Broadened Spectra and Its Temperature Dependence

One of the most important spectral parameters reflecting the electron-phonon interaction efficiency in solids is the homogeneous component of spectral line broadening δ.

For the majority of activated crystals the temperature range for investigating homogeneous broadening $\delta(T)$ has a lower limit corresponding to the inhomogeneous broadening level Δ: $\delta(T) > \Delta$. It is not possible to use traditional spectroscopic techniques in this temperature range to determine the homogeneous broadening component in disordered crystals and glasses. This is due to the very substantial magnitude of the inhomogeneous broadening component in these materials, which is of the same order of magnitude as Stark splitting ΔE and substantially exceeds homogeneous broadening $\Delta \approx \Delta E \gg \delta(T)$. However, selective laser excitation even in such unfavorable conditions permits direct measurement of homogeneous spectral broadening [16-18, 56, 57, 59, 60] and investigation of its temperature dependence.

We demonstrate below how the capabilities of selective spectroscopy for investigating the homogeneous broadening component of the inhomogeneously-broadened spectra of rare earth ions are implemented. The resonant scheme of selective laser excitation and recording at the same electron transition makes it possible to substantially limit the contribution of inhomogeneous broadening to the luminescence spectrum, reducing this contribution to a quantity comparable to the homogeneous component δ [46, 117].

We examine a variety of observed luminescence spectra with specific ratios of three spectral parameters: the spectral width of the excitation light Δv_e, the homogeneous linewidth of the electron transition δ and the inhomogeneous spectral width of the electron transition Δ. Here we will especially focus on the case of predominant inhomogeneous broadening, i.e., $\Delta \neq \delta$ which is characteristic of vitreous materials. Irradiation by light of spectral width Δv_e exceeding inhomogeneous broadening ($\Delta v_e \gg \Delta \gg \delta$ is nonselective excitation) produces an integral inhomogeneous luminescence spectrum of the resonant transition which is identical in shape to the inhomogeneous absorption spectrum $n(v)$.

The most important and valuable case from the viewpoint of useful data is the case of a narrow excitation light line when Δv_e is less than both parameters determining the spectral properties of the centers ($\Delta v_e \ll \delta \ll \Delta$). Here an inhomogeneous energy distribution of the centers arises on the excitation level [46, 117]

$$n^*(v, v_e) \sim \frac{\delta'}{(\delta')^2 + (v - v_{\mathrm{B}})^2} \, ,$$

having a temperature-dependent Lorentz contour of width δ'. In luminescence each optical center also radiates a Lorentz spectrum of identical width δ''. The integral luminescence spectrum of the specimen takes the form of the convolution of the two Lorentzian spectra (absorption and luminescence spectra) and also is a Lorentz spectrum

$$I(v, v_e) \sim \frac{\delta' + \delta''}{(\delta' + \delta'')^2 + (v - v_{\mathrm{B}})^2}$$

with a temperature-dependent width equal to the sum $\delta' + \delta''$. When luminescence is excited and observed at the same transition (the resonant scheme) the homogeneous absorption and luminescence widths are equal: $\delta' = \delta'' = \delta$. The luminescence spectrum in this case will have a Lorentz contour of width 2δ [46, 117] and a temperature dependence of the width analogous to the dependence for homogeneous broadened spectra.

The Sm^{3+} *ion* [131]. Figure 16a (curve 1) shows the collective luminescence spectrum corresponding to the transition between the lower Stark sublevels of the metastable state $^4G_{5/2}(1)$ and the ground state $^6H_{5/2}(1)$ of the Sm^{3+} ions in Li-La-phosphate glass under nonselective excitation. This spectrum yields information on the energy distribution of the centers $n(v)$ and is accurately described by a symmetrical Gaussian curve (characteristic of inhomogeneous broadening) of half-width $\Delta = 95$ cm^{-1} which is temperature-independent. This figure also provides the luminescence spectra measured under conditions of laser excitation at $\lambda_L = 5612$ Å in temporal strobing conditions ($t_d = 5 \cdot 10^{-5}$ s) at three different temperatures.

It is clear that all these spectra are substantially different both with respect to contour shape and width. Curve 4 is a Gaussian contour due entirely to the spectral resolution of the instrument Δ_{in} indicating low-level homogeneous broadening $\delta(T)$ of the resonant luminescence line of Sm^{3+} at 77 K ($\delta(T) \ll \Delta_{\mathrm{in}}$). Increasing temperature (curves 2, 3) results in a Lorentzian broadening component of the spectral lines, which becomes predominant at large T: $J_{\mathrm{in}} \ll \delta(T) \ll \Delta$.

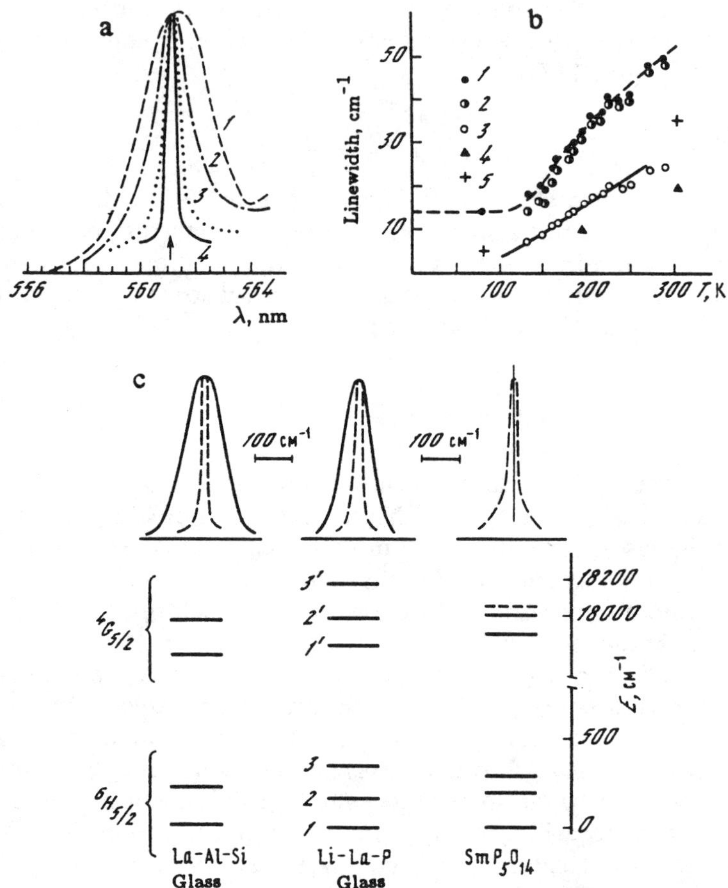

Fig. 16. Analysis of homogeneous broadening of the transitions and their temperature dependences in a number of solid-state Sm^{3+}-doped hosts.

a - luminescence contour of the Sm^{3+} ions ($^4G_{5/2}$—$^6H_{5/2}$) in Li-La-phosphate glass under nonselective (1) and selective laser excitation $\lambda_L = 5612$Å (2-4) (1 - lamp excitation, $T = 77$ K; 2-4 - laser excitation, $T = 300$, 137, and 77 K, respectively); b - temperature depen dences of the total linewidth (1), its Lorentz section (2) and the homogeneous component σ (3-5) for the $^4G_{5/2}$—$^6H_{5/2}$ transition (Sm^{3+}) under monochromatic excitation in Li-La-phosphate (1-3), and Li-Al-silicate (4) glasses and in an SmP$_5$-O$_{14}$ crystal (5); c - Stark splitting schemes of the $^4G_{5/2}$ and $^6H_{5/2}$ levels of Sm^{3+} in La-Al-silicate and Li-La-phosphate glasses and in an SmP$_5$O$_{15}$ crystal.

A line shape analysis demonstrated that the line shapes are accurately described by a Voigt contour which is a convolution of the Lorentz (or half-width 2δ) and Gaussian (of half-width Δ_{in}) lines at low temperatures. In the high temperature range the contour shape becomes purely Lorentzian. By isolating the Gaussian (instrument) component Δ_{in} from the Voigt contour we could draft the temperature dependence of Lorentzian broadening of the resonance transition ${}^{4}G_{5/2}(1) - {}^{6}H_{5/2}(1)$. Figure 16b provides the temperature dependences of the total measured linewidth (the dashed line and solid circles 1) and its Lorentzian part (the half-open circles 2). It should be pointed out that in the low temperature range where Lorentzian broadening is comparable to Gaussian broadening it is not possible to determine the homogeneous spectral line broadening component by simple subtraction of the Gaussian instrument function from the total measured width, as is occasionally done [44], as this will result in significant errors.

The open circles labeled 3 in Figure 16b represent the temperature dependence of the homogeneous width of the ${}^{4}G_{5/2}(1) - {}^{6}H_{5/2}(1)$ resonance transition obtained by dividing the Lorentzian linewidth of the spectra by a factor of 2.

The temperature rise of the homogeneous spectral line broadening component in the 77-300 K range is most often determined by the temperature increase of the probabilities of the nonradiative relaxation transitions between the Stark sublevels of the ground and excited states [59]. We have attempted to describe the experimental relation obtained for Li-La-phosphate glass as a single-phonon relaxation process, assuming that the primary contribution to broadening derives from the nonradiative transitions to the nearest Stark levels, which have a minimum energy gap $\Delta E_{1,2} \approx \Delta E_{1,2'} = 165$ cm^{-1} (see Figure 16c). The other Stark components are at a greater energy distance from the transition levels and therefore the nonradiative transitions at these levels can be neglected at moderately high temperatures.

In Figure 16b the solid line represents the relation

$$\delta_{1',1}(T) = 32[\exp(165/kT) - 1]^{-1}, \tag{16}$$

which corresponds to direct one-phonon relaxation transitions between the lower levels at the ground (1-2) and excited (1'-2') states with absorption of the vibrational energy radiation $\hbar\omega = 165$ cm^{-1}. It is clear that this relation accurately describes resonant line broadening, particularly in the moderately high temperature range. The suppressed experimental widths at $T \approx 300$ K may be due to the routine used to calculate these widths by the formula $\delta = \delta_{lor}/2$ which begins to lose validity as the homogeneous width δ approaches the inhomogeneous width Δ, since the coefficient 1/2 rises to 1 as $\delta \to \Delta$.

We carried out analogous measurements of the homogeneous broadening component of the transition between the lower component of the metastable level and the second component of the ground state $^4G_{5/2}(1)$ — $^6H_{5/2}(2)$ in Li-La-phosphate glass (see Figure 16c). The measurements were carried out at T = 147 K using the resonant scheme under conditions of selective excitation (λ_L = 567 nm) at the $^6H_{5/2}(2) - {}^4G_{5/2}(1')$ transition. The spectral luminescence line in this case was also accurately described by a Lorentz contour of width $2\delta_{1'2} \approx 90$ cm^{-1} which is smaller than the corresponding inhomogeneous broadening level of this transition $\Delta_{1'2} \approx 150$ cm^{-1} by a factor of approximately 1.7.

We interpret the substantial difference between the homogeneous broadening levels for the first δ_1, $\delta_{1'} \leqslant \delta_{1'1}$ = 8 cm^{-1} and second $\delta_2 = \delta \ll 1'2$ – d$_1 \approx$ 45-37 cm^{-1} Stark components as direct evidence of a relaxational broadening mechanism of the Sm^{3+} levels. Indeed the Raman broadening mechanism and similar Raman mechanisms [11, 103-109, 132, 133] are ordinarily quite insensitive to the relative position of the electron levels, are independent of the gap and will yield similar homogeneous broadening values, determined solely by the ratio between temperature and the effective phonon energy (the Debye case [100, 104], or the low-frequency disorder mode [132-134]). Unlike these broadening mechanisms in the relaxation mechanism the homogeneous width of the level is determined from the uncertainty relation based on the ronradiative transition rate for transitions between different Stark states: $\delta_i \approx \Sigma_k w_{ik}$ [135].

Here the ratio of the homogeneous widths of levels 1 and 2 is equal to the ratio of the nonradiative transition probabilities between these levels resulting in the emission or absorption of a phonon of energy equal to the energy gap ΔE_{12}: $\delta_1/\delta_2 = w_{12}/w_{21} = \exp(-\Delta E_{12}/kT)$.

For this case ΔE_{12} = 165 cm^{-1} and kT = 103 cm^{-1} (147 K) the w_{12}/w_{21} ratio is 0.2 which closely corresponds to the experimental width ratio of δ_{11}/δ_{12} = 0.18.

We compare the homogeneous widths and their temperature dependences for Sm^{3+} ions in Li-La-phosphate glass with similar values for the other vitreous and crystalline hosts. In Figure 16b the solid triangles labeled 4 represent the experimental homogeneous widths at the $^4G_{5/2}(1') \rightarrow {}^6H_{5/2}(1)$ test transition [76] for Sm^{3+} ions in La-Al-silicate glass at two different temperatures: δ = 9 cm^{-1} at 194 K and δ = 19 cm^{-1} at 300 K. The technique described above was used for the measurements. The crosses in this same figure represent the homogeneous widths of the $^4G_{5/2}(1')-{}^6H_{5/2}(1)$ transition in a SmP$_5$O$_{14}$ single crystal whose composition is similar to that of the Li-La-Sm-phosphate glass considered above. The nonselective luminescence contour of the SmP$_5$O$_{14}$ crystal was Lorentzian both at 300 K (δ = 34 cm^{-1}) and at 77 K (δ = 5 cm^{-1}), indicating minor spectral and structural disorder of the crystal: $\Delta \ll \delta$ (77 K) = 5 cm^{-1}

The homogeneous and inhomogeneous widths for these three activated media can be compared based on Figure 16c whose lower section contains the Stark splitting diagrams of the ground and metastable states. Analysis suggests that the homogeneous broadening levels and their variation with temperature for both the La-Al-silicate glass and the SmP_5O_{15} crystal are accurately described by the one-phonon nonradiative relaxation mechanism between the Stark sublevels:

$$\delta(T) = 20(e^{235/kT} - 1)^{-1} \text{ for La-Al-silicate glass}$$

$$\delta(T) = 23(e^{100/kT} - 1)^{-1} \text{ for } SmP_5O_{14}.$$

It is clear from Figure 16c that the phosphate glass has a lower Stark splitting level compared to the silicate glass: $\Delta E_{phos}/\Delta E_{sil} = 0.7$. The inhomogeneous broadening levels have a similar ratio: $\Delta_{phos}/\Delta_{sil} = 0.79$, indicating a lower absolute field strength of the ligands in Li-La-phosphate glass compared to La-Al-silicate glass with identical relative fluctuations: $\Delta_{phos}/\Delta E_{phos} \approx \Delta_{sil}/\Delta E_{sil} \approx 0.6$.

In summarizing these data we conclude that the attenuation of the ligand field in the switch from a silicate to phosphate host produces a simultaneous drop in the inhomogeneous broadening component Δ and a rise in the homogeneous component δ. The result is a precipitous change in the spectral inhomogeneity parameter ($\Delta/\delta = 6.3$ for the La-Al-silicate glass and $\Delta/\delta = 3.8$ for the Li-La-phosphate glass) which evidently explains the substantial difference in the behavior of phosphate and silicate laser glasses with rare earth ions in light amplifiers and oscillators. If we recall that the interaction of monochromatic light with a group of optical centers with inhomogeneous broadening results in a homogeneous component with a factor of 2 it becomes clear that phosphate glass can be considered to be a material with quasihomogeneous broadening, which differentiates it from silicate glass.

These results indicate that it is not necessary to incorporate concepts relating to the presence of low-frequency vibrational excitations (disorder modes) in the glass in order to explain the levels or temperature dependences of homogeneous broadening in the 100-300 K range in the case of Sm^{3+} ions in glass, as was done by the authors of Refs. [47, 48] to explain the anomalously high width $\delta(Eu^{3+})$ in glass compared to crystals.

The Eu^{3+}, Nd^{3+}, Yb^{3+} *ions.* We employed techniques analogous to those described above to measure the homogeneous components of the inhomogeneously-broadened bands for the Eu^{3+}, Nd^{3+}, and Yb^{3+} ions in a number of disordered media. The numerical values of these parameters are provided in Table 3 and are compared to the temperature, inhomogeneous broadening level Δ_I and the nonradiative transition energy ΔE_{ij}. The temperature relations $\delta(T)$ measured for Nd^{3+} and Yb^{3+} ions in phosphate glasses are shown in Figure 17.

It is clear from Figure 17 and Table 3 that all measured homogeneous widths and their temperature dependences closely correspond to the relaxational broadening mechanism resulting from one-phonon nonradiative transitions between the Stark sublevels. Significant changes in the force con-

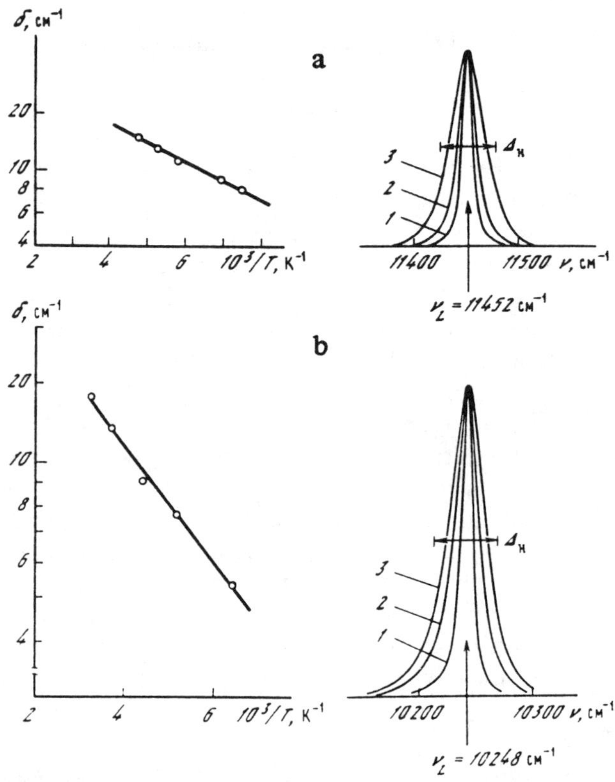

Fig. 17. The selective luminescence spectra at various temperatures and the $\delta(T)$ relations for the $^4F_{3/2}(1)$—$^4I_{9/2}(1)$ resonant transition of Nd^{3+} ions in Li-Al-phosphate glass (a) and the $^2F_{5/2}(1)$—$^2F_{7/2}(1)$ transition of Yb^{3+} ions in Ba-Al-phosphate glass (b).

a: 1 - T = 77 K, 2 - 118, 3 - 172; b: 1 - T = 77 K, 2 - 190, 3 - 300.

stant of the nonradiative transition are observed from one rare earth ion to another and when the crystal or vitreous matrices are substituted; such changes, however, make it impossible to trace the theoretically-predicted cubic dependence of the transition rate on transition energy: $\delta \sim w \sim \Delta E^3$ (see Expression (20), (21), and (24) from the second paper in the present volume). The data in Table 3 do not indicate levels substantially exceeding the broadening values for rare earth ions in glasses compared to crystals as observed in Refs [47, 48, 136] at lower temperatures, and made it necessary to introduce specific broadening mechanisms attributable to the dynamic structural disorder of glass to explain this fact [132–134, 137].

Table 3

Homogeneous broadening δ and the parameters A and ΔE_{ij} of its temperature dependence $\delta(T) = A(e^{\Delta E_{ij}/kT} - 1)^{-1}$ for rare earth ions in a number of crystals and glasses

Ion	Matrix	T, K	Δ_Γ, cm^{-1}	δ, см$^{-1}$	A, см$^{-1}$	$\Delta E_{i,j}$, см$^{-1}$	Ref.
Eu^{3+}	Na–Y–B–sili-cate glass	300	100	4.5 (1′–1)	6	180 (1–2)	[52]
		300	200	10 (1′–2)			
		300	200	25–30 (1′–3), 25–30(1′–4)	–	–	
Sm^{3+}	La–Al–sili-cate glass	194	120	9 (1′–1)	20	230(1–2)	[76]
		300	120	19 (1′–1)			
	Li–La–phos-phate glass	147	95	8 (1′–1)	32	165(1–2)	[131]
		147	150	45 (1′–2)	–	–	
	SmP$_5$O$_{14}$ crystal	77	< 5	5 (1′–1)	23	100(1′–2′)	[131]
		300	< 5	34 (1′–1)			
Nd^{3+}	CaF$_2$–YF$_3$ crystal	194	22	16 (1′–1)	–	–	[77]
		300	22	23 (1′–1)			
	Li–La–phos-phate glass	134–210	80	8–16 (1′–1)	12	84(1–2, 1′–2′)	[57]
Yb^{3+}	Ba–Al–phos-phate glass	118–300	55	1.0–15 (1′–1)	24	200(1–2, 1′–2′)	[57]
	Y$_3$Al$_5$O$_{12}$ crystal	80–250	0,8	0.3–17 (1′–1)	70	303(1′–2′)	[95]

Our experiments indicate that even Eu^{3+} ions which have a characteristic minimum homogeneous width have a significantly higher width for the transitions to the upper-lying $^7F_1$ (1, 2, 3) Stark levels compared to the lower $^7F_0$ level. Since we can neglect broadening of the metastable singlet state $^5D_0$ accurate to the very narrow natural width the ratio of the homogeneous spectral widths $\gamma_{1'1}(^5D_0-^7F_0)$ and $\delta_{1'2}(^5D_0-^7F_1(1))$ in an approximation of the relaxational broadening mechanism will correspond to the ratio of the nonradiative $^7F_0-^7F_1(1)$ transition rate resulting in absorption of a phonon $w^{\leftarrow} = w_0(e^{\Delta E/kT} - 1)^{-1}$ and the reverse nonradiative $^7F_1(1)-^7F_0$ transition resulting in emission of a phonon $w^{\rightarrow} = w_0(1 - e^{\Delta E/kT})^{-1}$. As we see the ratio $\vec{w}\vec{w} = e^{-\Delta E/kT}$ also determines the minimum (relaxational) value of the homogeneous widths:

$$\frac{\delta_{1'1}(^5D_0-{}^7F_0)}{\delta_{1'2}(^5D_0-{}^7F_1(1))} = e^{-\Delta E/kT}$$

For the experimental conditions using Eu^{3+} ions ($\Delta E_{12} = 180$ cm^{-1} and $T = 300$ K) the factor $\exp(-\Delta E/kT)$ yields a value of 0.43 which is in good agreement with the measured ratio $\delta_{1'1}(^5D_0-{}^7F_0)/\delta_{1'2}(^5D_0-{}^7F_1(1)) = 0.45$ (see Table 3).

We believe that an elevated ratio of the homogeneous width of the lower level to the upper level compared to the calculated relaxation factor is the only serious evidence in favor of an auxiliary (aside from relaxational) broadening mechanism of the Raman or other combination type.

A predominance of the latter mechanisms can therefore be expected to occur with greater probabilities at low temperatures ($T < 100$ K, $kT \ll \Delta E$, see results [47, 48, 136]) when the low population of the phonon states of energies of the order of ΔE will result in a low probability of nonradiative relaxation transitions with phonon absorption ($\hbar\omega = \Delta E$), and not at high temperatures (see results [137–139]) when the similarity of the temperature relations to an exponential law becomes evident due only to the limited accuracy of the experiments or as a result of the random complex multilevel nature of the relaxational nonradiative transitions.

7. Dispersion of the Metastable State Decay Rates and the Radiative and Nonradiative Transition Probabilities in Optical Centers with Different Spectral Properties

In addition to spectral line broadening and splitting, relaxation properties also play an important role in the spectroscopy of activated media. Such properties include the lifetime τ_i, and the radiative (A_i) and nonradiative (w_i) transition probabilities at different optical centers. The integral luminescence decay kinetics of specimens is therefore quite complex for many glasses and crystals containing rare earth ions when ordinary nonselective [14] excitation is used and the kinetics cannot be approximated by a single exponent. It is quite important to identify the structure of this non-exponential property and to understand the processes behind it, since such events bear information on the relaxation channels of electron excitations in laser and luminescent media.

The Nd^{3+} *ion.* Figures 18a and c show sample results from an investigation of the elementary relaxation characteristics of isolated groups of

Nd^{3+} optical centers in LGS-28 silicate glass [56, 78] by selective laser excitation. Tuning of the excitation light serves to alter the luminescence decay rate (see Figure 18a). Figure 18c provides the dispersion curve of the average lifetimes $\bar{\tau}(\lambda_L)$ obtained by integrating the nonexponential decay kinetics $\bar{\tau}(\lambda) = I_0^{-1}\int_0^\infty I(\lambda, t)dt$ for each excitation wavelength. It is clear that $\theta(\lambda_L)$ varies by a factor exceeding 2 from one group of optical centers to another. The observed changes may be due to either the dispersion of the radiative transition probabilities from the varying "forbiddance" on the electrical dipole positions at the different optical centers or the dispersion of the nonradiative transition probabilities occurring as a result of the differential between the vibronic interaction forces at optical centers in the glass. An analysis of the nonstationary spectra of monochromatic excitation ($^4I_{9/2} - {}^2G_{7/2}, {}^4G_{5/2}$) from nonselective recording of the integral luminescence of Nd^{3+} centers [56, 78] yields an answer to the question of which of these mechanisms is the determinant mechanism.

Figure 18b shows two nonstationary laser excitation spectra $I(\lambda_L)$ recorded in the initial ($t_d = 10^{-5}$ s $\ll \bar{\tau}$) and final ($t_d = 3 \cdot 10^{-3} \gg \bar{\tau}$) stages of the decay kinetics (curves 1 and 2, respectively). This figure also shows the integral absorption spectrum of this transition (curve 3). During the recording process the $I(t_d, \eta_L)$ spectra were automatically corrected for the change in laser excitation power $P(\eta_L)$ by dividing the two signals: $I(t_d, \lambda_L)/P(\lambda_L)$. The existence of intensive spectrum 2 indicates that the inhomogeneous absorption contour contains the longer-lived Nd^{3+} centers whose luminescence, unlike the short-lived centers, does not decay 3 ms after the pulsed excitation, which correlates well with the relation $\bar{\tau}(\lambda_L)$ shown in Figure 18c.

Since nonstationary spectrum 2 is similar in shape to spectrum 5 in Figure 9a it is possible to identify the position of these long-lived centers at the $^4F_{3/2}-{}^4I_{9/2}$ luminescent transition (see center 5 in Figure 9b). The excitation spectrum recorded with a short delay $t_d \ll \bar{\tau}$ (Figure 18b, curve 1) also was similar in shape to the integral absorption spectrum. However, the intensity distribution in the first spectrum bears additional information on the dispersion of the radiative probabilities $A(\lambda_L)$ of the $^4F_{3/2}-{}^4I_{9/2}$ luminescent transition. Indeed, if the absorption spectrum $k(\lambda_L) = \sigma(\lambda_L)n(\lambda_L)$ reflects only the number of centers $n(\lambda_L)$ and the capacity $\sigma(\lambda_L)$ of the optical center to absorb a quantum of excitation light, the luminescence excitation spectrum will include, in addition to $k(\lambda_L)$, the capacity $A(\lambda_L)$ to emit the absorbed energy at the luminescent transition in the case of a low optical density of the specimen: $I(0, \lambda_L) \sim P(\lambda_L)k(\lambda_L)A(\lambda_L)$. Dividing the intensity of the excitation spectrum by the intensity of the absorption spectrum provides us with the dispersion characteristic $A(\lambda_L)$ shown in Figure 18c.

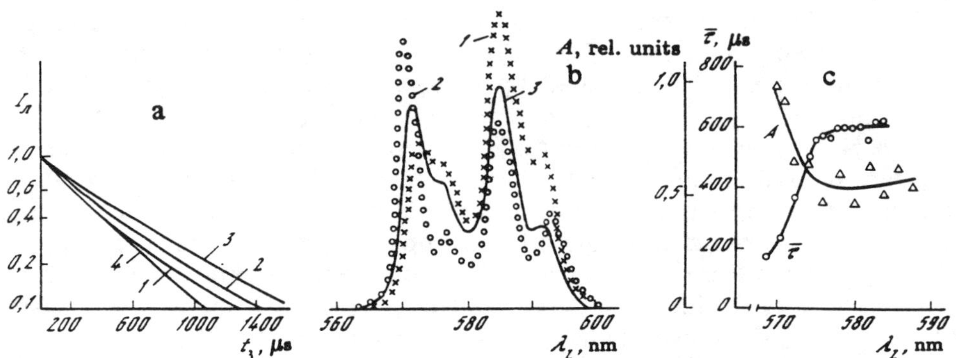

Fig. 18. Results of an analysis of the dispersion of the lifetime and transition probabilities from the $^4F_{3/2}$ level of Nd^{3+} ions in LGS-28 glass under conditions of selective laser excitation.

a - luminescence decay kinetics from the $^4F_{3/2}$ level of Nd^{3+} ions for η_L = 566 (1), 567 (2), 568 (3) and 573 nm (4) for T = 4.2 K; b - luminescence excitation spectra of Nd^{3+} in LGS-28 glass (1, 2) and the absorption spectrum (3) at the $^4I_{9/2}$-$^2G_{7/2}$, $^4G_{5/2}$ transition (1 - t_d = 10 μs, 2 - t_d = 3 ms); c - average luminescence decay time τ from the $^4F_{3/2}$ level and the $^4F_{3/2}$-$^4I_{9/2}$ radiative transition probability A plotted as a function of the wavelength of the excitation light λ_L for T = 77 K.

It is clear from the figure that the dispersion of the radiative transition probabilities of the Nd^{3+} ions correlates accurately with the dispersion of the luminescence decay lifetimes $\bar{\tau}(\lambda_L)$ in LGS-28 glass. This explains the nonexponential nature of the luminescence kinetics of the test laser medium with varying "forbiddance" on the electrical dipole transitions at nonequivalent optical centers of the glass. The authors of Ref. [93] employed our technique to investigate phosphate, tellurite, germanate, and silicate (ED-2) glass with Nd^{3+}. They identified a significant dependence of τ on λ_L for the germanate matrix only.

The Yb^{3+} *ion.* We also observed correlated behavior of $\bar{\tau}(\nu)$ and $A(\nu)$ in the case of Yb^{3+} ions in Ba-Al-phosphate glass (Figure 19). The technique differed in the method used to measure the $A(\nu)$ relation. In this case we compared the absorption spectrum of the Yb^{3+} ions at the $^2F_{7/2}(1)$—$^2F_{5/2}(1)$ transition between the lower Stark components (curve 1 in Figure 19a) to the luminescence spectrum of the same transition excited by means of a pulsed broadband $LiF(F_2 - F_2^+)$ laser with $\Delta\nu$ = 200 cm^{-1} (curve 2 in Figure 19a). For the resonant experimental setup used in this case this made it possible to eliminate continuous tuning as well as correction of the spectral dependence of the laser output power, which was required in the case of Nd^{3+} ions. Dividing the intensity of curve 2 by the intensity of

curve 1 yielded the frequency dependence of the radiative transition probability $A(v)$ (see Figure 19a).

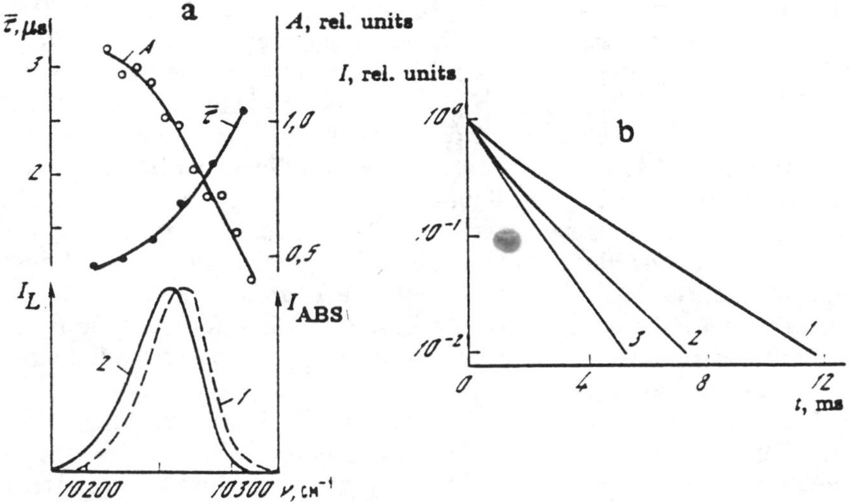

Fig. 19. Variations in the average lifetime $\bar{\tau}$ and spontaneous radiative probability A of the $^2F_{5/2}(1)$—$^2F_{7/2}(1)$ transition (a) and in the luminescence
decay kinetics $I_L(t)$ (b) with variations in the selective laser excitation frequency for Yb^{3+} ions in Ba–Al–phosphate glass at T = 4.2 K.
b: 1 - v_L = 10,309, 2 = 10,266, 3 = 10,224 cm^{-1}.

We carried out measurements of the luminescence decay kinetics of
Yb^{3+} ions using double selection by means of selective narrowband excitation and recording at the same frequency ($v_L = v_r$) which was varied in
discrete steps within the homogeneous contour. The average values of τ in
Figure 19a were obtained in the same manner as the Nd^{3+} ions based on the
decay kinetics at different frequencies (see Figure 19b). The correlated
variations in the $\bar{\tau}(v)$ and $A(v)$ ions as in the case of Nd^{3+} ions indicates a
proportionality of $\bar{\tau}(v) \sim A^{-1}(v)$ and a low efficiency of the intracenter
nonradiative relaxation channel in these laser matrices.

Similar studies carried out for the same purpose on neodymium- and
yttrium-containing glasses using traditional spectroscopic techniques differ-

ing from ours (i.e., lower spectral and temporal resolution than the laser case) were carried out in Refs. [99, 141-144]. The authors observed significant changes in luminescence decay kinetics as a function of excitation and recording frequencies on the inhomogeneous contour. However, these studies did not (with the exception of Ref. [144]) carry out an independent analysis of the dispersions of $\bar{\tau}(v)$ and $A(v)$ and postulated that $A(v) = \bar{\tau}^{-1}(v)$. Here the nonradiative transition probability w and its dispersion $w(v)$ were neglected, which is by no means justified in all cases.

The authors of Refs. [88, 89] in investigating the relaxational properties of Nd^{3+} ions in glasses developed their own technique using selective laser excitation to the $^2P_{1/2}$ singlet level. They identified some interesting excitation frequency dependences of the luminescence branching coefficients $I_L(^4F_{3/2}-^4I_{11/2})/I_L(^4F_{3/2}-^4I_{9/2})$, and the total and radiative deactivation rates of the $^4F_{3/2}$ level of Nd^{3+} ions in silicate, phosphate, borate, fluoroberyllium, and fluorophosphate glasses. On the other hand they devoted significant attention to identifying the contribution of the nonradiative transitions, although their sharpest $w(v)$ relations (calculated as $w(v) = \bar{\tau}^{-1}(v) - A(v)$) were found for glasses where the differences between $\bar{\tau}(v)^{-1}$ and $A(v)$ were not substantial (10-25%) and could be caused by methodological or random errors. In the single case of borate glass where the nonradiative transition rate is reliably determined ($w(v) \gg A(v)$) it is nearly independent of the excitation frequency of the $^4I_{9/2} - {}^2P_{1/2}$ transition: $w(v) \sim$ const.

The Cr^{3+} ion. The (3d) transition group ions may have different dispersions of the transition probabilities compared to rare earth ions. In the absence of optical electron screening, average or strong crystal field and strong electron-phonon coupling conditions are realized [68, 103, 145-154]. This is manifested as substantial inhomogeneous and quasi-inhomogeneous broadening of their multiphonon spectra [154, 155]. Selective excitation and recording proved to be effective in investigating such complex systems (see Ref. [121] as well as section 4 of the present paper).

Ref. [120] was devoted to selective spectroscopy of Mo^{3+} ions (or the $4d^3$ configuration) in phosphate glass; this study identified strong vibronic coupling, inhomogeneous broadening and variations of the luminescence decay time at the $^2E-^4A_2$ transition which is normally insensitive to the crystal field. An investigation of the $4f^n - 15d - 4f$ allowed transitions of the Ce^{3+} and Eu^{2+} ions in quartz glasses carried out in Ref. [123] also identified a strong dependence of the lifetime on the excitation wavelength. Here we examine the properties of Cr^{3+} ions in La-Al-borate glass [118, 119].

The spectroscopic properties of Cr^{3+} ions in glasses include a low position of the $^4T_2$ level which becomes a metastable level, and a precipitous nonexponential nature of the decay kinetics as well as a strong temperature dependence of the lifetime [154, 155] due to the low intracrystal field parameters (D_q). The spectroscopic properties of La–Al–borate glass with Cr^{3+} ions include a high radiative transition probability and a short radiative lifetime. Our figure, based on the absorption spectrum, was $\bar{\tau}_0 = \bar{A}^{-1} = 12$ μs and was independent of temperature over the range 77–300 K. Measurements of the integral luminescence decay kinetics at 77 K yielded a value of $\bar{\tau} = I_0^- \int_0^\infty I(t)dt = 9$ μs for the average lifetime. This indicates a high luminescence quantum efficiency of the $^4T_2$ level in this material ($\eta = \bar{\tau}/\bar{\tau}_0 = 0.75$) which differentiates it from other media (such as phosphate glasses where $\bar{\tau}_0$ is an order of magnitude greater with identical values of $\bar{\tau}$).

The continuing nonexponential nature of the decay kinetics at $T = 4.2$–77 K (Figure 20a) where $A > w$ indicates substantial dispersion of τ_0 and A at the various optical centers. In this case the frequency shift of the low-temperature luminescence spectra recorded at different times t_d (Figure 20b) and the variation in the luminescence decay kinetics in the case of selective recording (see Figure 20a) indicate that the radiative probability drops for centers with a high transition energy. A rise in temperature to 300 K causes the average lifetime $\bar{\tau}$ to drop to 4 μs and the quantum efficiency η to drop to 0.3. However, we note that the latter remains substantially higher compared to other vitreous materials [154–156].

Temperature luminescence quenching at 300 K resulting from the predominance of the nonradiative excitation decay channel ($w > A$ and $\eta = A/(A + w) < 0.5$) makes it possible to determine the dispersion of the non-radiative probabilities $w(v)$ based on measurements of the decay kinetics at 300 K. It is clear from Figure 20a (curves 3 and 4) that lower decay rates at all kinetic stages correspond to centers with a higher transition energy ($^4T_2$–$^4A_2$). Such a reduction in w with increasing $E(^4T_2$–$^4A_2)$ obviously is related to the higher activation energy of the nonradiative transition ($\epsilon_I > \epsilon_{II}$) for the high-frequency centers $E_I > E_{II}$ ($w(T) \sim \exp(-\epsilon/kT)$), i.e., to the higher position of the point of intersection of the radiative level $^4T_2$ and the ground level $^4A_2$ from the minimum of the $^4T_2$ potential curve (see Figure 20c).

Such a simple treatment based on the generally-accepted nonradiative transition model for the case of strong electron–phonon coupling [103] will, as we shall see, successfully explain the observed kinetic and spectral characteristics of excitation relaxation of both Cr^{3+} and Mo^{3+} ions [120] in various glasses and we therefore believe that it is not appropriate to introduce more complex and less well-justified models [156] for this purpose.

It is interesting to compare the spectral dependences of the radiative transition probabilities $A(v)$ for the Eu^{3+}, Yb^{3+}, Nd^{3+} and Cr^{3+} ion glasses to their electron transition energies. Thus, for the Nd^{3+}, Yb^{3+}, and Cr^{3+} ions it was determined that the radiative electric dipole transition probabilities (due to the large odd crystal field parameters) are greater for centers with small even crystal field parameters (B_n^m, D_q) and therefore with lower Stark splitting energies (Nd^{3+}, Yb^{3+}) or transition energies (Cr^{3+}). This suggests that there is a rise in symmetry (and a reduction in distortion) in glasses of smaller size (higher density) of the coordination polyhedron of NdO_8, YbO_6, CrO_6. An analysis of the literature data [89, 93, 141-144] confirms this conclusion for a broad range of silicate, borate, fluorophosphate, and germanium glasses containing Nd^{3+} and for silicate and phosphate glasses containing Yb^{3+} which makes it possible to apply this method to a broad range of glasses containing rare earth ions.

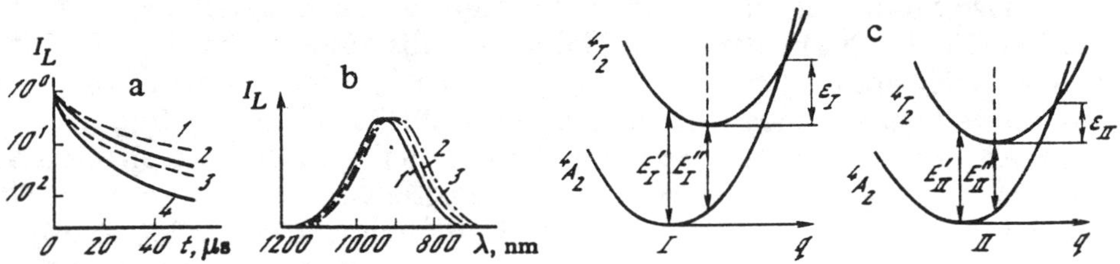

Fig. 20. Kinetic (a) and spectral (b) manifestations of the dispersion of the radiative $A(\lambda)$ and nonradiative $W(\lambda)$ transition probabilities and their correlation with the $^4T_2-^4A_2$ luminescence transition energy (c) for Cr^{3+} ions in La-Al-borate glass.

a - deactivation kinetics of the $^4T_2$-level under selective excitation ($\lambda_r = 0.53$ μm) and luminescence recording (1, 3 - $\lambda_r = 0.8$ μm, 2, 4 = $\lambda_r = 0.92$ μm, 1, 2 - $T = 77$ K; 3, 4 - $T = 300$ K); b - the luminescence spectra at various points in the decay kinetics following "instantaneous" excitation (t_d, μs: 1 - 1.5, 2 - 40, 3 - 80); c - models of the Cr^{3+} electron centers with various absorption E' and luminescent E'' transition energies and their relation to the nonradiative transition activation energy ϵ.

A certain correlation of A and ΔE is also observed for Eu^{3+} ions in Li-La- and Ba-Al-phosphate glasses, although it is reversible in nature and is substantially less evident (a dispersion of the lifetime $\bar{\tau}(v)$ is not found). This can be understood as a weak rise in the electric dipole probability of the forbidden transitions from $^5D_0$ to the $^7F_0$ and $^7F_1$ levels due to their mixing with the $^7F_2$ level, which is enhanced for centers with larger intracrystal field parameters B_n^m. Studies of the luminescence decay kinetics of Sm^{3+} ions in La-Al-silicate and Li-La-phosphate glasses under selective laser excitation as in the case of Nd^{3+} ions in Li-La-phosphate glass did not reveal a significant dispersion of $\bar{\tau}(v)$, which accurately correlates with the near-exponential integral luminescence decay in these media. A comparison of the nonstationary excitation spectra of Sm^{3+} in Li-La-phosphate glass recorded with both short ($t_d \ll \bar{\tau}$) and long ($t_d \gg \bar{\tau}$) delays to the absorption spectrum did not reveal differences indicating the existence of any significant dispersion of the radiative probabilities $A(v)$ at the $^4G_{5/2}(1)-^6H_{5/2}(1)$ transition.

Conclusion

In this paper we have considered the application of both existing and new selective laser spectroscopic techniques to structural analysis of the electron and vibronic states, radiative and nonradiative relaxation channels, and ion-lattice interaction mechanisms in disordered, condensed matter with inhomogeneously-broadened spectra of rare earth ions. This made it possible to identify the structure of inhomogeneous broadening, detect and investigate the details of the Stark splitting of the inhomogeneously-broadened absorption and luminescence spectra of the activator ions and to draft composite Stark sublevel diagrams for continuous sets of optical centers in glass.

In addition to the spectra of the electron transitions we also identified a number of vibronic spectra, thereby demonstrating the exponential nature of the frequency dependence of the vibrational state density over the range 20-200 cm^{-1}. The magnitudes and temperature dependences of homogeneous broadening of the inhomogeneously-broadened spectra of activated glasses and crystals indicate the predominance of the relaxational broadening mechanism resulting from one-phonon inter-Stark nonradiative transitions.

The significant dispersion of the intracenter radiative and nonradiative transition probabilities identified for a number of glasses is in satisfactory correlation with the variation in the electron transition and Stark splitting energy from one optical center to another, which made it possible to identify the mechanism behind the increasing symmetry of the coordinational polyhedron of the impurity ion with diminishing ion-ligand distance.

It is important to note that the studies outlined above were carried out on crystals and glasses activated with Nd^{3+}, Cr^{3+}, Yb^{3+}, Sm^{3+} and Eu^{3+} ions that not only function as convenient model systems but also have found broad practical application as laser and luminescent media in quantum electronics.

References

1. A. A. Kaminskiy, and V. V. Osiko, "Inorganic Laser Materials with a Fine Structure", *Bull. Acad. Sci., USSR*, vol. 1, pp. 2049-2087, vol. 3, pp. 417-463; 1970, vol. 6, pp. 629-696, 1965 (in Russian).
2. A. A. Kaminskiy, *Lazernye kristally* [Laser Crystals], Moscow: Nauka, 1975, 255 pp (in Russian).
3. G. O. Karapetyan and A. L. Reyshakhrit, "Luminescent Glasses as a Material for Optical Quantum Oscillators", *Bull. Acad. Sci. USSR*, vol. 3, 1967, pp. 217-259.
4. N. E. Alekseev, V. P. Gapontsev, and M. E. Zhabotinskiy, et al., *Lazernye stekla* [Laser Phosphate Glasses] Moscow: Nauka, 1980, 352 pp (in Russian).

5.   Yu. V. Denisov and V. A. Kizel', "Energy Migration in Europium-Activated Borate Glasses and the Relative Location of Energy Levels", *Opt. and Spectros.* vol. 23, no. 3, 1967, pp. 472-474.

6.   Yu. V. Denisov, B. F. Dzhurinskiy, and V. A. Kizel', "Investigation of Luminescence and Energy Migration in Glass-Like Rare-Earth Solutions", Bull. Acad. Sci. USSR, Phys. Ser., vol. 32, no. 9, 1968, pp. 1580-1583 (in Russian).

7.   Yu. V. Denisov, B. F. Dzhirinskiy, and V. A. Kizel', "Inhomogeneous and Homogeneous Broadening of Rare-Earth Element Lines in Glasses", In: *Inhomogeneous Broadening of the Spectral Lines of Laser Active Media*, (IFAN Ukr. SSR, Kiev), 1969, pp. 209-216 (in Russian).

8.   I. V. Basil'ev, G. M. Zverev, G. Ya. Kolodnyi, and A. M. Onishchenko, "Nonresonance Excitation Energy Transfer Between Impurity Rare-Earth Ions", *Sov. Phys. JETP*, vol. 56, no. 1, 1969, pp. 122-123 (in Russian).

9.   A. Szabo, "Laser-Induced Fluorescence Line Narrowing in Ruby", *Phys. Rev. Lett.*, vol. 25, no. 14, 1970, pp. 924-926.

10. A. Szabo, "Observation of the Optical Analog of Mössbauer Effect in Ruby", *Ibid.*, vol. 27, no. 6, 1971, pp. 323-326.

11. L. A. Riseberg, "Observation of Laser Induced Fluorescence Line Narrowing in Glass-Nd", *Ibid.*, vol. 28, no. 13, 1972, pp. 786-789.

12. L. A. Riseberg, "Temperature Dependence of Ion-Ion Energy Transfer in Nd^{3+} Glass by Laser Induced Line Narrowing Techniques", *Solid State Communs.*, 1972, pp. 469-471.

13. L. A. Riseberg, "Laser Induced Fluorescence Line Narrowing Spectroscopy of Glass-Nd", *Phys. Rev. A. Gen. Phys.*, vol. 7, no. 2, 1973, pp. 671-678.

14. R. I. Personov, E. I. Al'shits, and L. A. Bykovskaya, "The Effect of Line Structure Appearance in Laser-Excited Fluorescence Spectra of Organic Compounds in Solid Solutions", *Opt. Communs.*, vol. 6, 1972, pp. 169-173.

15. R. I. Personov, E. I. Al'shits, L. A. Bykovskaya, and B. M. Kharlamov, "The Fine Structure of the Luminescence Spectra of Organic Molecules Under Laser Excitation and the Nature of the Broad Spectral Bands of Solid Solutions", *Sov. Phys. JETP*, vol. 65, no. 5, 1973, pp. 1825-1836 (in Russian).

16. K. K. Perane, R. A. Avarmaa, and A. A. Gorokhovskiy, "The Structure of Zero-Phonon Lines in the Inhomogeneously-Broadened Luminescence Spectra Under Monochromatic Excitation", *Bull. Acad. Sci. USSR*, vol. 39, no. 9, 1975, pp. 1793-1800 (in Russian).

17. R. I. Personov, "Nature of the 'Smeared-out' Bands in the Electronic Spectra of Organic Compound Solutions and Methods of Detecting its Latent Line Structure", *Ibid.*, vol. 42, no. 2, 1978, pp. 242-252.

18. R. I. Personov, *Selective Spectroscopy of Complex Molecules in Solutions and its Application*, Preprint of the Spectroscopy Institute, Acad. Sci. USSR, no. 14, Troitsk, 1981, 24 pp.

19. A. G. Cabesas and R. P. Treat, "Effect of Spectral Hole Burning and Cross-Relaxation on the Gain Saturation of Laser Amplifiers", *J. Appl. Phys.*, vol. 37, no. 9, 1966, pp. 3556-3563.

20. V. R. Belan, I. M. Briskina, V. V. Grigor'yants, and M. L. Guraru, "Investigation the Nature of Laser Luminescence Spectral Line Broadening", In: *Inhomogeneous Spectral Line Broadening of Laser Active Media*, Kiev: IFAN USSR, 1969, pp. 151-160 (in Russian).

21. V. R. Belan, I. M. Briskana, V. V. Grigor'yants, and M. E. Zharotinskiy, "Excitation Energy Transfer Between Neodymium Ions in Glass", *Sov. Phys. JETP*, vol. 57, no. 4, 1969, pp. 1148-1159 (in Russian).

22. V. V. Grigoryants, "Information Content of Luminescence of the Active Medium of a Free-Running Laser", *Sov. Phys. JETP.*, vol. 58, no. 5, 1970, pp. 1593-1605 (in Russian).

23. V. V. Grigoryants, *Luminescence of Laser Media Under Coherent Irradiation*, Dissertation for Doctor Phys.-Math, Moscow, 1971, 305 pp.

24. V. V. Grigoryants, M. E. Zhabotinski, and V. B. Markuchev, "Lasing and Spectral Line Characteristics in Phosphate Glasses and Inorganic Liquid with Neodymium", *IEEE J. Quant. Electron.*, vol. 8, no. 2, 1972, pp. 196-198.

25. V. I. Nikitin, M. S. Soskin, and A. I. Khizhnyak, "New Results on the 1.06 μm Luminescent Band Structure of Nd^{3+} Ion in Silicate Glass", *JETP Lett.*, vol. 2, no. 4, 1976, pp. 172-176 (in Russian).

26. V. I. Nikitin, M. S. Soskin, and A. I. Khizhnyak, "Uncorrelated Inhomogeneous Broadening: The Principal Cause of Narrowband Lasing in ND^{3+} Phosphate Glass", *Ibid.*, vol. 3, no. 1, 1977, pp. 14-18.

27. V. I. Nikitin, M. S. Soskin, and I. A. Khizhnyak, "Effect of Uncorrelated Inhomogeneous Broadening of the Nd^{3+} 1.06 μ Band on Laser Properties of Neodymium Glass", *Sov. Jour. Quantum Electronics*, vol. 5, no. 6, 1978, pp. 1375-1379 (in Russian).

28. V. I. Nikitin, M. S. Soskin, and A. I. Khizhnyak, "Deformation of the 1.06 μ Nd^{3+} Band Profile of Glasses in Free-Oscillation Conditions", *Ibid.*, vol. 6, no. 10, 1979, pp. 2400-2443.

29. B. M. Kharlamov, R. . Personov, and L. A. Bykovskaya, "Stable "Gap" in Absorption Spectra of Solid Solutions of Organic Molecules by Laser Irradiation", *Opt. Communs.*, vol. 12, 1974, p. 191.

30. A. A. Gorokhovskiy, R. A. Avartmaa, and K. K. Rebane, "Hole Burning in the Contour of a Pure Electronic Line in the Shpol'skii System", *JETP Lett.,*, vol. 20, no. 7, 1974, p. 474 (in Russian).

31.   A. R. Gutierrez, J. Friedrich, D. Haarer, and H. Wolfrum, H., "Multiple Photochemical Hole Burning in Organic Glasses and Polymers Spectroscopy and Storage Aspects", *IBM J. Res. and Develop*, vol. 26, no. 2, 1982, pp. 198-208.

32.   V. P. Lebedev and A. K. Prezhevuskiy, "Polarized Luminescence of Glasses Activated by Rare-Earth Ions", *Sov. Phys. Solid State*,1977, vol. 19, no. 8, 1977, pp. 1373-1376 (in Russian).

33.   A. K. Prezhevuskiy, "Inhomogeneous Spectral Structure of Glasses Activated by Ions of Rare-Earth Elements", In: *Spect. of Crystals*, Leningrad: Nauka, 1978, pp. 96-108 (in Russian).

34.   V. P. Lebedev and A. K. Prezhevuskiy, "Determination of Multipolarities of Optical Transitions in the Spectra of Glasses Activated by Eu^{3+} Ions by the Polarized Luminescence Technique", *Opt. and Spectroscopy,* vol. 48, no. 5, 1980, pp. 932-935 (in Russian).

35.   B. Ya. Zabokritskiy, A. D. Manuil'skiy, and M. S. Soskin, "Investigation of the Nd^{3+} Energy Spectrum ni Crystals and Glasses Under Narrowband Excitation", In: *Abstracts of III Repub. Seminar on Quantum Electronics*, Khar'kov, 1973, p. 14 (in Russian).

36.   N. Montegi and S. Shionoya, "Excitation Migration Among Inhomogeneous Broadened Level of EU^{3+} Ions", *J. Luminescence*, vol. 8, 1973, pp. 1-17.

37.   M. Y. Yen, S. S. Sussman, J. A. Paisner, and M. J. Weber, "Fluorescence Line Narrowing and Energy Migration in Up Converted Transition EU^{3+} Glass", *Lawrence Livermore Lab. Prepr. UCRL-76481,* 1975, pp. 1-13.

38.   C. Brecher and L. A. Riseberg, "Laser-Induced Fluorescence Line Narrowing in Eu^{3+} Glass: A Spectroscopic Analysis of Coordination Structure", *Phys. Rev. B - Solid State*, vol. 13, no. 1, 1979, pp. 81-93.

39.   M. Y. Weber, J. A. Paisner, S. S. Sussman, et al., "Spectroscopic Studies of Rare-Earth Ions in Glass Using Fluorescence Line Narrowing Techniques", *J. Luminescence*, no. 12/13, 1976, pp. 729-735.

40.   C. Delsart, N. Pelletier-Allard, and R. Pelletier, "Affinement d'une raie de fluorescence de Pr^{3+}:$LaAlO_3$ par exitation laser", *Opt. Communs.*, vol. 11, no. 1, 1974, pp. 84-88 (in French).

41.   L. E. Erickson, "Fluorescence Line Narrowing of Trivalent Praseodymium in Lanthanum Trifluoride Single Crystal. Phonon-Induced Relaxation", *Phys. Rev. B - Solid State*, vol. 11, 1975, pp. 77-81.

42.   P. M. Selser, D. S. Hamilton, R. Flach, and W. M. Yen, "Phonon-Assisted Energy Migration in Pr^{3+}:LaF_3", *J. Luminescence*, no. 12/13, 1975, pp. 737-741.

43.   R. Flach, D. S. Hamilton, P. M. Selser, and W. M. Yet, "Time-Resolved Fluorescence Line-Narrowing Studies in LaF_3:Pr^{3+}", *Phys. Rev. Lett.*, vol. 35, no. 15, 1975, pp. 1034-1037.

44. Ch. Hsu and R. C. Powell, "Dye Laser Spectroscopy of Trivalent Samarium in Calcium Tungsten Crystals", *J. Phys. C: Solid State Phys.*, vol. 9, 1976. pp. 2467-2480.

45. Yu. V. Denisov, I. V. Kovaleva, V. P. Kolovkov, and V. V. Rastoskuev, "Structure of the Environment and Luminescence Line Broadening of Europium in Glass-Like Matrices", *Opt. and Spectroscopy*, vol. 38, no. 1, 1975, pp. 98-103 (in Russian).

46. T. Kushida and E. Takushi, "Determination of Homogeneous Spectral Widths by Fluorescence Line Narrowing in $Ca(PO_3)_2$:Eu^{3+}", *Phys. Rev. B - Solid State*, vol. 12, no. 3, 1975, pp. 824-827.

47. P. M. Selzer, D. L. Huber, D. S. Hamilton, et al., "Observation of Anomalous Fluorescence Line Width Behavior in Amorphous Materials Using Fluorescence Line Narrowing", *Proc. Conf. Struct. Amorph. Solids*, Williamsburg, VA, 1976, p. 5.

48. P. M. Slezer, D. L. Huber, D. S. Hamilton, et al., "Anomalous Fluorescence Line Width Behavior in Eu^{3+}-Doped Silicate Glass", *Phys. Rev. Lett.*, vol. 36, no. 14, 1976, pp. 813-816.

49. T. Kushida, E. Takushi, and Y. Oka, "Memories of Photon Energy Polarization and Phase in Luminescence of Rare-Earth Ions Under Resonant Light Excitation", *J. Luminescence*, no. 12/13, 1976, pp. 723-727.

50. T. T. Basiev, *Electronic Excitation Transfer Between Rare-Earth Ions in Laser Matrices*, Ph. D. Dissertation, Phys. and Math., Moscow: FIAN SSSR, 1976 (in Russian)

51. T. T. Basiev, "Assembly for Spectroscopic Investigations of Solids by the Oscillographic Delayed-Coincidence method" *Sov. Pribory i tekh. Eks.*, no. 2, 1976, pp. 182-185 (in Russian).

52. O. K. Alimov, T. T. Basiev, Yu. K. Boron'ko, et al., "Laser Spectroscopy of Inhomogeneously-Broadened Lines in Glasses and Migration of Electronic Excitation", *Sov. Phys. JETP*, vol. 72, no. 4, 1977, pp. 1313-1327.

53. O. K. Alimov, T. T. Basiev, and Yu. K. Boron'ko, "Luminescence and Energy Migration in Silicate Glass Activated by Eu^{3+} Ions Under Monochromatic Laser Excitation", In: *Abstracts of the VII All-Union Conf. on Coherent and Nonlinear Optics*, Tbilisi: Metsniereva, vol. 1, 1976, p. 63 (in Russian).

54. O. K. Alimov, T. T. Basiev, and Yu. K. Boron'ko, "Laser Spectroscopy of the Inhomogeneously-Broadened Lines of Nd^{3+} Ions in Laser Crystals and Glasses", In: *Abstracts of the First All-Union Conf. on Laser Optics*, Leningrad: GOI, 1976, pp. 15-16 (in Russian).

55. O. K. Alimov, T. T. Basiev, Yu. K. Boron'ko, and A. V. Dmitryuk, "Luminescence and Energy Migration in Silicate Glass, Activated by Eu^{3+} Ions Under Monochromatic Laser Excitation", In: *Abstracts of*

the V All-Union Symposium on Spectroscopy of Crystals, Activated by Rare-Earth Ions, Kazan: University of Kazan, 1976, p. 13.

56. T. T. Basiev, Yu. K. Borno'ko, and A. M. Prokhorov, "Structural Investigation of the Inhomogeneously-Broadened Spectra of Rare-Earth Ions and Energy Migration Inhomogeneous Contour by Selective Laser Excitation", In: *Spectroscopy of Crystals*, Ed. by P. P. Feofilov, Leningrad: Nauka, 1978, pp. 83-96 (in Russian)..

57. T. T. Basiev, Yu. K. Boron'ko, V. V. Osiko, and A. M. Prokhorov, A. M., "Laser Spectroscopy of Activated Crystals and Glasses, In: *Spectroscopy of Crystals*, Ed. by A. A. Kaplyanskii, Leningrad: Nauka, p. 57 (in Russian).

58. P. M. Selzer, "General Techniques and Experimental Methods in Laser Spectroscopy of Solids", *Laser Spectroscopy of Solids*, Ed. by. W. M. Yen and P. M. Selzer, Berlin, etc.: Springer, 1981, pp. 113-140.

59. W. M. Yen and P. M. Selzer, "High Resolution Laser Spectroscopy of Ions in Crystals", *Ibid.*, pp. 141-188.

60. M. J. Weber, "Laser Excited Fluorescence Spectroscopy in Glass", *Ibid.*, pp. 189-239.

61. G. B. Abdullaev, T. T. Basiev, Yu. K. Voron'ko, et al., "Pulsed-Periodic Differential Frequency Oscillations in GaSe with Continuous Tuning over the Range 5.5-18 μm", In: *Abstracts of the All-Union Conf. on Coherent and Nonlinear Optics*, Kiev, part 1, pp. 258-259 (in Russian).

62. O. K. Alimov, M. Kh. Ashurov, and T. T. Basiev, T. T., *Kinetic Luminescent Spectroscopy of the Rare-Earth Ions in Crystals and Glasses Under Selective Laser Excitation*, Moscow: Preprint FIAN No. 172, 1980, 24 pp.

63. A. M. Bonch-Bruevich, *Radioelektronika v eksperimental'noy fizike*, [Radioelectronics in Experimental Physics], Moscow: Nauka, 1966, 752 pp. (in Russian).

64. T. T. Basiev, Yu. I. Malakhov, and S. K. Tsys', S. K., "Lifetime of the $^2\Sigma_u^+CO_2^+$ Level" *Opt. and Spectroscopy*, vol. 30, 1971, pp. 421-424.

65. T. T. Basiev, Yu. I. Malakhov, and V. G. Potemkin, "Measurement of the Lifetimes of Some A_rI Levels by the Oscillographic Delayed-Coincidence Method", In: *Research on Elementary Processes in Low-Temperature Plasma*, Cheboksary: Univ. of Cheboksary, 1973, pp. 3-6 (in Russian).

66. Yu. K. Voron'ko, A. A. Kaminskiy, and V. V. Osiko, "Analysis of the Optical Spectra of Pr^{3+}, Nd^{3+}, Eu^{3+}, Er^{3+} in Fluorite Crystals (Type 1) by the Concentration Series Method", *Sov. Phys. JETP*, vol. 49, no. 3, 1965, pp. 724-729.

67. M. I. Gauduk, V. F. Zolin, L. S. Gaygerova, *Spektry Lyuminestsentsii Evropiya* [Luminescence Spectra of Europium], Moscow: Nauka, 1978, 195 pp.

68. D. T., Sviridov, R. K. Sviridova, Yu. F. Smirnov, *Opticheskie Spektry Ionov Perekhodnykh Metallov v Kristallakh* [Optical Spectra of Transitional Metals Ions in Crystals] Moscow: Nauka, 1976, 266 pp.

69. E. F. Kustov, G. A. Bandurkin, E. N. Murav'ev, and V. P. Orlovskiy, *Elektronnye Spektry Soedineniy Redkozemel'nykh Elementov* [Electronic Spectra of Rare-Earth Element Compounds], Moscow: Nauka, 1981, 303 pp.

70. M. J. Weber, J. Hegarty, and D. H. Blackburg, "Laser-Induced Fluorescence Line Narrowing of Eu^{3+} in Lithium Borate Glass", *Lawrence Livermore Lab. Prepr. UCRL-79530*, 1977, p. 11.

71. J. Hegarty, W. M. Yen, M. J. Weber, and D. H. Blackburg, "Laser-Induced Fluorescence Line Narrowing of Eu^{3+} in Lithium Borate Glass", *J. Luminescence*, no. 18/19, 1979, pp. 657-660.

72. C. Brecher and L. A. Riseberg, "Laser-Induced Line Narrowing of Eu^{3+} Fluorescence in Fluoberyllate Glass: Site-Dependent Spectroscopic Properties and Their Structural Implications", *Phys. Rev. B - Solid State*, vol. 21, no. 6, 1980, pp. 2607-2618.

73. M. J. Weber, "Fluorescence Line-Narrowing Studies in Glass, *Spectroscopie des èlèments lourds des solides*, Lyon, 1977, pp. 283-287. (Colloq. intern. CNRS; N 255).

74. V. I. Aleksandrov, M. A. Borik, G. Kh. Dechev, et al. "Synthesis and Investigation of Glass in the La_2O_3-Al_2O_3-SiO_3 System" *Fizika i Khimiya Stekla*, vol. 6, no. 2, 1980, pp. 170-173.

75. V. M. Ivanova and S. S. Denisova, "Glass", *Proceedings of the Scientific Research Institute of Glasses*, no. 1(140), 1971, p. 81 (in Russian).

76. T. T. Basiev, M. A. Borik, Yu. K. Voron'ko, et al., "Selective Laser Excitation of Luminescence of Sm^{3+} Ions in Lanthanoalumosilicate Glass", *Opt. and Spectrosc*, vol. 46, no. 5, 1979, p. 510.

77. T. T. Basiev, Yu. K. Borno'ko, A. Ya. Karasik, et al., "Spectral Migration of Electronic Excitation Between Nd^{3+} Ions in CaF_2-YF_3 Crystals Under Selective laser Excitation Conditions", *Sov. Phys. JETP*, vol. 75, no. 1(7), 1978, p. 32.

78. O. K. Alimov, T. T. Basiev, Yu. K. Voron'ko, et al., "Selective-Laser-Excitation Study of the Structure of Inhomogeneously-Broadened Spectra of Nd^{3+} Ions in Glass", *Ibid.*, vol. 47, p. 29.

79. T. T. Basiev, Yu. K. Boron'ko, S. B. Mirov, and A. M. Prokhorov, "Frequency Selection of Nd^{3+} Ions in Glass, Excited by Monochromatic Laser Radiation at the Resonant Transition $^4I_{9/2}$-$^4F_{3/2}$", *JETP Lett.,*, vol. 29, 1979, p. 639.

60   *Alimov et al.*

80.   O. K. Alimov, T. T. Basiev, and Yu. K. Boron'ko, "Stark Structure of the $^4I_{9/2}$ State of the Nd^{3+} Ion in Phosphate Glass Under Selective Laser Excitation", *Sov. Phys. Lebedev Institute Reports*, no. 9, 1980, pp. 22-25.

81.   O. K. Alimov, T. T. Basiev, Yu. K. Boron'ko, and V. S. Fedorov, "A New Method of Laser-Spectroscopy Investigation of Solids with Inhomogeneously-Broadened Spectra", *JETP Lett.*, vol. 29, 1979, p. 127.

82.   O. K. Alimov, T. T. Basiev, and Yu. K. Boron'ko, "Investigation of the Fine Structure of Inhomogeneously-Broadened Absorption Bands of the Nd^{3+} Ion in Glass by the Selective Excitation Method", *Sov. J. Quant. Electr*, vol. 10, 1980, p. 1443.

83.   T. T. Basiev, Yu. K. Voron'ko, S. B. Mirov, et al., "Kinetics of the Accumulation and Excitation of F_2^+-Centers in $LiF(F_2)$ Crystals", *Sov. JETP Lett.*, vol. 30, 1979, p. 626.

84.   T. T. Basiev, S. B. Mirov, and A. M. Prokhorov, "Pulsed Periodic Tunable Laser Using an LiF Crystal with F_2^+-Centers Excited by Second Harmonic Radiation from an Nd^{3+} Garnet Laser", *Proc. Acad. Sci. USSR*, vol. 246, 1979, pp. 374.

85.   R. I. Personov and E. T. Al'shits, "A New Possibility for Determining Vibration Frequencies of Organic Molecules in the Excited Electronic States from Fluorescence Spectra", *Chem. Phys. Lett.*, vol. 33, no. 1, 1975, pp. 85-88.

86.   R. I. Personov and B. M. Kharlamov, "Extreme Narrowing of Bands in the Fluorescence Excitation Spectra of Organic Molecules in Solid Solution", *Opt. Communs.*, vol. 7, no. 4, 1973, pp. 417-419.

87.   J. Hegarty, W. M. Yen, M. J. Weber, "Spectroscopic Properties of Excited Ions in Glass Measured Using Laser-Induced Fluorescence Line Narrowing", *Phys. Rev. B - Solid State*, vol. 18, no. 10, 1978, pp. 5816-5819.

88.   M. J. Weber, C. Breher, and L. A. Riseberg, "Variation in the Transition Probabilities and Quantum Efficiency of Nd^{3+} Ions in ED-2 Laser Glass", *Appl. Phys. Lett.*, vol. 35, no. 1, 1979, pp. 31-33.

89.   C. Brecher, L. A. Riseberg, and M. J. Weber, "Line-Narrowed Fluorescence Spectra and Site-Dependent Transition Probabilities of Nd^{3+} in Oxide and Fluoride Glasses", *Phys. Rev. B - Solid State*, vol. 18, no. 10, 1978, pp. 5799-5811.

90.   S. A. Brawer and M. J. Weber, "Observation of Fluorescence Line Narrowing, Hole Burning, and Ion-Ion Energy Transfer in Neodymium Laser Glass", *Appl. Phys. Lett.*, vol. 35, no. 1, 1979, pp. 31-33.

91.   M. V. Glushkov, Yu. V. Kosichkin, V. V. Osiko, et al., "Discrimination of Inhomogeneously-Broadened Emission Spectra of Neodymium by Resonance Laser Excitation", *Sov. J. Quant. Electr.*, vol. 9, 1976, p. 1296.

92. V. I. Nikitin, M. S. Soskin, and A. I. Kizhnyak, "Luminescence of Neodymium Glasses Under Narrowband Excitation in the Resonant Transition Range $^4I_{9/2}$-$^4F_{3/2}$" *Ukrain. Fiz. Zh.*, vol. 25, no. 9, 1979, pp. 1543-1548.

93. Y. Kalisky, R. Reisfeld, Y. Haas, "Spectral Behavior of Nd^{3+} Doped Glasses Under Narrow-Line Excitation", *Chem. Phys. Lett.*, vol. 61, no. 1, 1979, pp. 19-22.

94. *Spectroscopy of Crystals*, Leningrad: Nauka, 1973, 331 pp.; 1975, 383 pp.; 1978, 192 pp.; 1983, 230 pp. (in Russian).

95. T. T. Basiev, Yu. K. Voron'ko, and I. A. Shcherbakov, "Effect of Spectral Line Broadening on the Migration of Electronic Excitation Along Crystal Impurity Centers", *Sov. Phys. JETP*, vol. 39, 1974, p. 1042.

96. T. T. Basiev, Yu. K. Voron'ko, T. G. Mamedov, and I. A. Shcherbakov, "Energy Migration Between Yb^{3+} Ions in Garnet Crystals", *Sov. J. Quant. Electr.*, vol. 5, 1975, p. 1182.

97. T. T. Basiev, Yu. K. Voron'ko, T. G. Mamedov, et al., "Relaxation Excitation Processes of Metastable Levels of Rare-Earth Ions in Crystals", In: *Spectroscopy of Crystals*, Ed. by A. A. Kaminskii, Z. A. Morgenshtern, and D. T. Sviridov, Moscow: Nauka, 1975, p. 155 (in Russian).

98. P. P. Feofilov, "Collective Optical Phenomena in Activated Crystal" In: *Physics of Impurity Centers in Crystals*, Proc. of Intern. Seminar "Select Problems in the Theory of Impurity Centers in Crystals", Tallin, 1972, 539 pp. (in Russian).

99. V. P. Lebedev, A. K. Przhevuskii, and V. A. Svost'yanov, "Spectral and Kinetic Selection of Optical Centers in Glass", *Fiz. i Khim. Stekla*, vol. 5, 1979, p. 608 (in Russian).

100. A. Maradudin, *Defects and the Vibrational Spectrum of Crystals*, Moscow: Mir, 1968, 432 pp. (in Russian).

101. K. K. Rebane, *Elementary Theory of the Vibrational Spectral Structure of Impurity Centers in Crystals*, Moscow: Nauka, 1968, 232 pp. (in Russian).

102. I. V. Ignatyev and V. V. Ovsyankin, "Vibronic Spectra and the Dynamics of Crystals with Impurity Rare-Earth Ions", In: *Spectroscopy of Crystals*, Ed. by A. A. Kaplyanskii, Leningrad: Nauka, 1983, p. 36 (in Russian).

103. Yu. E. Perlin, B. S. Tsukerblatt, *Vibronic Interaction Effects in the Optical Spectra of Paramagnetic Impurity Ions*, Kishinev: Shtiintsa, 1974, 368 pp. (in Russian).

104. I. S. Osad'ko, "Study of Vibronic Interaction, Based on the Structure of the Optical Spectra of Impurity Centers", *Sov. Phys. Uspekhi*, vol. 128, no. 1, 1979, pp. 31-67 (in Russian).

106. M. A. Krivoglaz, "Theory of Phonon-Free Line Broadening in the Mössbauer or Optical Spectrum", *Sov. Phys. Solid State*, vol. 6, 1964, p. 1707.

107. D. E. McCumber, "Width of the Narrow No-Phonon Lines on the Optical Spectra of Impurities in Solids", *Phys. Rev. A., Gen. Phys.*, vol. 133, no. 1, 1964, pp. 166-168.

108. I. S. Osad'ko, "The Theory of Temperature Broadening and Shift of Phonon-free Optical Impurity Lines" *Sov. Phys. Solid State*, vol. 14, 1972, p. 2927.

109. I. S. Osad'ko, "Zero-Phonon Lines and Phonon Wings in the Absorption and Luminescence Spectra of Impurities", *Ibid.*, vol. 17, p. 3180, 1975.

110. R. Avarmaa, "Effect of Monochromatic Excitation on the Width and Intensity of Inhomogeneously-Broadened Lines in the Luminescence Spectrum", *Izv. AN EstSSR, Phys., Math.*, vol. 23, no. 3, 1974, p. 238-247 (in Russian).

111. Ja. Kikas, "Determination of the Inhomogeneous Distribution Function and Homogeneous Component of Impurity Molecular Spectra", *Ibid.*, vol. 25, no. 4, 1976, pp. 374-379 (in Russian)

112. Ja. Kikas, "Determination of Homogeneous Impurity Spectra in Strongly Inhomogeneous Solids", *Ibid.*, vol. 28, no. 1, 1979, pp. 73-75.

113. V. A. Savostyanov, A. K. Przhevuskii, "Energy Gap Dependence of Excitation Migration Along a Nonuniform Contour in Glass" *Sov. Phys. Solid State.*, vol. 21, 1979, p. 465.

114. E. N. Alekseev, V. P. Gapontsev, M. E. Zhabotinskii, and Yu. E. Sverchkov, "Measurement of Nonresonant Interactions of Rare-Earth Ions in Condensed Media by Selective Observation of Luminescence Kinetics", *JETP Lett.*, vol. 27, p. 109, 1978.

115. A. K. Przhevuskii, "Excitation Migration Among the Activator Centers in Crystals and Glasses", *Bull. Acad. Sci. USSR, Phys. Ser.*, vol. 45, p. 4, 1981.

116. O. K. Alimov, T. T. Basiev, and Yu. K. Voron'ko, "Phonon Distribution Function Study in Structurally-Disordered Solids by Monochromatic Excitation of Luminescence" *Sov. Phys. Solid State*, vol. 24, 1982, p. 1700.

117. I. I. Abram, R. A. Auerbach, R. R. Birge, et al., "Narrow-Line Fluorescence Spectra of Perylene as a Function of Excitation Wavelength" *J. Chem. Phys.*, vol. 63, 1975, p. 2473.

118. T. T. Basiev, M. A. Borik, Yu. K. Voron'ko, et al., "Laser Spectroscopy of Vibronic Transitions of Cr^{3+} Ions in Lanthan-Aluminum-Borate Glass", In: *Abstracts of the X Urals Symposium on Spectroscopy*, Sverdlovsk, 1980, pp. 70-71 (in Russian).

119. O. K. Alimov, T. T. Basiev, E. O. Kirpichenkova, S. B. Mirov, "Kinetic Luminescence Spectroscopy of Nd^{3+}, Yb^{3+}, and Cr^{3+} Ions in Solids Under Selective Laser Excitation" In: *Abstracts of the All Union Vavilov Symp. on Luminescence*, Leningrad, GOI, 1981, p. 30 (in Russian).

120. M. J. Weber, S. A. Brawer, and A. J. DeGroot, "Site-Dependent Decay Rates and Fluorescence Line Narrowing of Mo^{3+} in Phosphate Glass", *Phys. Rev. B, Solid State*, vol. 22, 1980, p. 1163.

121. S. A. Brawer, W. B. White, "Optical Properties of Trivalent Chromium in Silicate Glasses: A Study of Energy Levels in the Crossing Region", *J. Chem. Phys.*, vol. 67, 1977, pp. 2043-2055.

122. G. Boulon, B. Moine, J. C. Bourcet, "Spectroscopic Properties of $^3P_1$ and $^3P_0$ Excited States of Bi^{3+} Ions in Germanate Glass", *Phys. Rev. B, Solid State*, vol. 22, 1980, p. 1163.

123. V. I. Arbuzov, V. A. Bonch-Bruevich, E. I. Galant, et al., "Inhomogeneous Spectral Structure of Eu^{2+} and Ce^{2+} Ions in Quartz Glass", *Fiz. i Khimiya Stekla*, vol. 8, 1982, pp. 216-222 (in Russian).

124. R. Beigang, G. Litfin, and H. Welling, "Frequency Behavior and Line-Width of CW Single Mode Color Center Laser" *Opt. Communs.*, vol. 22, no. 3, 1977, pp. 269-271.

125. R. Beigang, G. Litfin, and H. Welling, "Color Centers for High Resolution Infrared Spectroscopy" *Tenth Intern. Quantum Electron. Conf.*, Atlanta, GA, 1978, May 29-June 21.

126. G. Litfin and H. Welling, "Color Center Lasers" *Nelineynaya Optika* [Nonlinear Optics], Novosibirsk: ITF SO AN SSSR, Chapter 1, 1979, pp. 250-256.

127. L. F. Molenauer and D. H. Olson, "Broadly Tunable Lasers Using Color Centers", *J. Appl. Phys.*, vol. 46, no. 7, 1975, pp. 3109-3118.

128. I. A. Parfianovich, V. M. Khulugurov, B. D. Lobanov, and N. T. Maksimova, "Lumnescence and Stimulated Emission of Color Centers in LiF", *Bull. Acad. Sci. USSR, Phys. Ser*, vol. 43, 1979, p. 20.

129. N. P. Ermikova, N. I. Ivanov, A. A. Mihalenko, et al., "LiF-Based Active Media With Color Centers for Oscillations in the 0.7-0.92 μm Spectral Range", In: *Abstracts of VII Repub. Conf. Junior Scientists "Investigations in Spectroscopy and Quantum Electronics"*, Vilnyus, 1983, p. 114 (in Russian).

130. V. M. Hulugurov and B. D. Lobanov, "Lasing at the Color Centers in LiF-OH at 300 K in the Spectral Range 0.84-1.13 μm, *Sov. Tech. Phys. Lett.*, vol. 4, 1978, p. 595.

131. A. G. Avanesov, T. T. Basiev, Yu. K. Voron'ko, et al., "Selective Laser Excitation Study of Electron-Phonon Interaction of Sm^{3+} Ions in Glass", *Opt. and Spectrosc.*, vol. 51, 1981, pp. 148-153.

64   *Alimov et al.*

132. S. K. Lyo and R. Orbach, "Homogeneous Fluorescence Line-Width for Amorphous Hosts", *Phys. Rev. B, Sold State*, vol. 22, no. 9, 1980, pp. 4223–4225.
133. I. S. Osad'ko, "Anomalous Thermal Broadening of the Optical Lines of Impurity Centers in Glass", *JETP Lett.*, vol. 33, 1981, p. 626.
134. T. L. Reineche, "Fluorescence Line-Width in Glasses", *Solid State Communs.*, vol. 32, no. 11, 1979, pp. 1103–1106.
135. W. M. Yen, W. C. Scott, and A. L. Shawlow, "Phonon-Induced Relaxation in Excited Optical States of Trivalent Pr^{3+} in LaF_3", *Phys. Rev. A, Gen. Phys.*, vol. 136, no. 1, 1964, pp. 271–283.
136. Y. Hegarty and W. M. Yen, "Optical Homogeneous Line-Width of Pr^{3+} in BeF_2 and GeO_2 Glasses", *Phys. Rev. Lett.*, vol. 43, no. 15, 1979, pp. 1126–1130.
137. Yu. V. Denisov, N. F. Perevozchikov, and A. P. Shubin, "Vibronic Interaction and Homogeneous Broadening of Europium Lines in Glass-Like Matrices", In: *Abstracts of the All-Union Symp. on Luminescence*, Leningrad: GOI, 1981, p. 49 (in Russian).
138. P. Avouris, A. Campion, and M. A. El-Sayed, "Variations in Homogeneous Fluorescence Line-Width and Electron Phonon Coupling Within an Inhomogeneous Spectral Profile", *J. Chem. Phys.*, vol. 67, no. 7, 1977, pp. 3397–3398.
139. N. F. Perevozchikov, *Manifestations of Structural-Dynamic Properties of a Matrix and the Interaction of Europium Ions in Activated Glass Spectra*, PhD Dissertation in Phys. and Math., Dolgoprudnyi, MFTI, 1982, 190 pp. (in Russian).
140. T. T. Basiev, E. M. Dianov, A. M. Prokhorov, and I. A. Shcherbakov, "On the Quantum Efficiency of Luminescence from the Metastable Level of Nd^{3+} in $Y_3Al_5O_{12}$ Silicate Glasses and Crystals", *Proc. Acad. Sci. USSR*, Dokl. Acad. Nauk SSSR, vol. 216, 1974, pp. 297–299.
141. A. R. Kangro, Ya. E. Karris, A. K. Przhevuskii, et al., "Variation in the Luminescence Spectra of Neodymium Glass in the Decay Process", *Sov. Tech. Phys. Lett.*, vol. 2, 1976, pp. 652–655.
142. V. I. Arbuzov, A. K. Przhevuskii, and M. N. Tolstoi, "Kinetic Selection of Excitation Spectra of Neodymium Glass", *Fiz. i Khimya Stekla*, vol. 3, 1977, p. 236 (in Russian).
143. A. K. Przhevuskii, V. A. Savost'yanov, and M. N. Tolstoi, "Chrono-spectroscopic Investigation of the Luminescence of Neodymium Glasses", *Sov. J. Quant. Electr.*, vol. 8, 1978, p. 54.
144. V. A. Savost'yanov, V. A. Malyshev, A. K. Przhevuskii, and A. S. Troshin, "Chronospectroscopic Study of Luminescence-Band Structure in Yb^{3+} Ions in Glass at Low Temperatures", *Opt. and Spectrosc.*, vol. 47, 1979, pp. 532–540.

145. J. C. Walling, H. R. Jenssen, R. C. Morris, et al., "Tunable-Laser Performance in BeAl$_2$O$_4$:Cr^{3+}" *Opt. Lett.*, vol. 4, no. 6, 1979, pp. 182-183.

146. J. C. Walling, O. G. Peterson, H. P. Jenssen, et al., "Tunable Alexandrite Lasers" *IEEE J. Quant. Electron.*, vol. 16, no. 12, 1980, pp. 1302-1314.

147. J. C. Walling, "Alexandrite Lasers: Physics and Performance" *Laser Focus*, Feb. 1982, pp. 45-50.

148. B. K. Sevast'yanov, Yu. L. Remigailo, V. P. Orekhova, et al., "Lasing Frequency Tuning and Spectroscopic Characteristics of Alexandrite", *Bull. Acad. Sci. USSR, Phys. Ser.*, vol. 40, 1981, p. 61.

149. B. Struve, G. Huber, V. V. Laptev, et al., "Laser Action and Broadband Fluorescence in Cr^{3+}-GdScGa-Garnet" *XII Intern. Quant. Electron. Conf.*, 1982, Munchen, pp. 235-236.

150. E. V. Zharikov, N. N. Il'ichev, S. P. Kalitin, et al., *Tunable Gadolinium-Scandium-Gallium, Garnet Crystal Laser Operating at the Vibronic Transition of Chromium*, Preprint No. 20, FIAN SSSR, Moscow, 1983, 6 pp. (in Russian).

151. M. L. Shand and J. C. Walling, "Tunable Emerald Laser" *IEEE J. Quant. Electron.*, vol. 18, no. 11, 1982, pp. 1829-1830.

152. H. P. Christensen, H. P. Jenssen, "Broad-Band Emission from Chromium Doped Germanium Garnets", *Ibid.*, no. 8, pp. 1197-1201.

153. P. T. Kenyon, L. Andrews, B. McCollum, and A. Lempicki, "Tunable Infrared Solid State Laser Materials Based on Cr^{3+} in Low Ligand Fields", *Ibid.*, pp. 1189-1196.

154. L. J. Andrews, A. Lempicki, and B. C. McCollum, "Spectroscopy and Photokinetics of Chromium (III) in Glass", *J. Chem. Phys*, vol. 74, no. 10, 1981, pp. 5526-5538.

155. G. A. Mokeeva, A. H. Hudoleev, N. M. Bokin, et al., "Luminescence of Trivalent Chromium in Phosphate and Germanate Glasses", In: *Spectroscopy of Crystals,* Ed. by P. P. Feofilov, Leningrad: Nauka, 1973, pp. 308-312 (in Russian).

156. A. G. Avanesov, A. I. Burshtein, B. I. Denker, et al., "Variance of the Probabilities of Intracenter Nonradiative Transitions in Solids", *Proc. Acad. Sci. USSR*, Dokl. Acad. Nauk SSSR, vol. 254, 1980, p. 737.

ELECTRON-EXCITATION ENERGY TRANSFER AMONG IMPURITY IONS IN DISORDERED MEDIA

O. K. Alimov, M. Kh. Ashurov, T. T. Basiev,
E. O. Kirpichenkova, V. B. Murav'ev

Abstract: **Theoretical models of quenching and migration nonradiative energy transfer for pair and collective particle interaction are discussed. Results are given from statistical computer modeling of the energy migration process in a disordered particle ensemble; such results have made it possible to identify the diffusion stage of the migration kinetics for dipole-dipole, dipole-quadrupole, quadrupole-quadrupole, and short-range interactions.**

Experimental techniques of nonstationary kinetic spectroscopy under selective laser excitation employing Yb^{3+}, Eu^{3+}, Sm^{3+}, and Nd^{3+} ions in laser glasses were used to investigate ion-ion nonradiative energy transfer, and electron excitation quenching and migration processes in the coordinate and frequency spaces for the case of chaotic particles and inhomogeneous broadening of the particle spectra. Quenching and migration interactions with elevated multipolarity are detected together with previously-unknown kinetic stages of excitation quenching and migration in disordered media. Optical techniques are used for the first time to identify the near order parameters, specifically the minimum distances of impurities in glasses.

Multistage resonant and single-stage nonresonant spectral energy migration over an inhomogeneously-broadened contour are investigated. A new method of investigating the dependence of energy transfer efficiency on the frequency of the phonon emitted in the process is proposed, substantiated and implemented. The average optical excitation quenching and migration rates and the efficiency of elementary interactions of active ions in disordered vitreous matrices are determined. The diffusion stage of excitation migration in a disordered particle ensemble is detected and investigated and the diffusion constants and their concentration dependence are determined.

The development of experimental nonstationary selective spectroscopic techniques and the analysis of the energy and relaxation properties of elementary optical centers (see the first paper in the present volume) makes it possible to proceed with an investigation of more complex and interesting problems of the condensed state, specifically, an analysis of collective interaction of impurity particles, which will be the subject of the present paper.

The problem of excitation transfer between impurity particles in a solid is a classical problem for solid state and luminescence physics [1, 2]. Aside from the fundamental scientific significance studies of optical energy transfer processes between absorption and emission have an important practical application. They determine the techniques used to fabricate new optical, luminescent and laser materials of enhanced efficiency and previously-unattainable properties [3-8].

The issues of excitation transfer between rare-earth ions have been the subject of research for a relatively long period (see surveys [9-12]) although such research focused largely on the qualitative and semiphenomenological levels. Only in the early 1970s did efforts focus on the active development of effective techniques for spectral-kinetic investigations of crystals containing rare-earth ions [7, 13-18] and kinetic theories of collective interactions between impurities were rapidly developed [1, 2, 19-28]. This made it possible to investigate the macroprocesses of electron excitation energy migration and nonradiative degradation in crystals and to obtain information on the ion-ion interaction mechanisms on the microlevel [7, 15, 17, 23]. In spite of the detailed development of such a kinetic approach the full collection of experimental confirmations accumulated to date remains far from complete and is limited solely to the dipole-dipole particle interaction mechanism.

In recent years researchers have focused extensive attention on doped glasses (see surveys [5, 6, 8, 29-32]) as valuable laser media and convenient model objects with a disordered structure. Unlike ordered crystal matrices it is substantially more difficult to achieve the microparameter level in research in disordered impurity media. The development of spectrally-selective kinetic techniques has provided a solution to this problem and also made it possible to search out and investigate interactions of higher multipolarity by focusing substantial attention on an analysis of the features of the spatial distribution of impurities in disordered solids.

1. Energy Transfer in an Interacting Ion Pair
Analysis of Theoretical Models

In this section we analyze the mechanisms and nature of ion-ion interaction in the simplest model of two optically-active centers. The probability of the elementary act of interaction W_{ij} represents the basis of any collective energy transfer processes such as donor-donor energy migration, sensitization, donor-acceptor quenching transfer, nonlinear summing of excitations, etc. The direct determination of the primary microparameters of this interaction: the efficiency C_{ij} and multipolarity S and their functional relations to the atomic and crystal constants ($W_{ij} = C_{ij}/R^S$, where R is the distance between particles) is an important problem of any research on energy transfer, since this makes it possible to establish the nature and mechanisms behind the interaction of active ions and to predict the behavior of actual macroscopic systems based on tested theoretical models.

The fundamental basis of the simplest and most natural approach to the transfer problem is first order perturbation theory which permits calculation of the probability of energy transfer from one ion (the donor) to another (the acceptor) [33-35]:

$$W_{DA} = \frac{2\pi}{\hbar} |\langle \psi_D' \psi_A | H_{DA} | \psi_D \psi_A' \rangle|^2 \rho(E). \tag{1}$$

Here $\langle H_{DA} \rangle$ is the matrix element of the transition from the initial state $\psi_D' \psi_A$, where the donor-ion is excited and the acceptor is not, to the final state $\psi_D \psi_A'$, where the acceptor is excited and the donor is not; H_{DA} is the Coulomb or exchange interaction between ions at a distance R (in the simplest case of dipole-dipole interaction $H_{DA} \sim R^{-3}$); $\delta(E)$ is the density of the final states per unit of energy.

Förster [33, 34] and Dexter [35] have demonstrated that in a number of cases Expression (1) can be written in a more convenient form through the experimentally-determined parameters of the donor and acceptor particles such as the oscillator strengths, lifetimes, or transition cross-sections as well as the emission and absorption spectra. For dipole-dipole interaction it takes the form

$$W_{DA} = \frac{9c^4}{8\pi \tau_D R^6} \int F_D(\nu) \sigma_A(\nu) \nu^{-4} d\nu, \tag{2}$$

where $\sigma_A(\nu)$ is the absorption cross-section of the acceptor transitions (in cm^2); $F_D(\nu)$ is the normalized emission spectrum of the donor under the normalization condition $\int F_D(\nu)d\nu = 1$; R is the distance between ions; τ_D is the donor radiative lifetime; c is the speed of light. It is clear that the rate of energy transfer grows with diminishing distance between the ions, with increasing cross-sections of the energy transitions D and A and with increasing spectral overlap integrals.

When the higher order interaction multipolarities (such as dipole-quadrupole or quadrupole-dipole $- H_{DA} \sim R^{-4}$ or quadrupole-quadrupole $- H_{DA} \sim R^{-5}$ interactions) make the predominant contribution to the interaction Hamiltonian, the expressions for the transfer probability W_{DA} are similar with different coefficients: R^6 and ν^{-4} are substituted with R^8 and ν^{-6} or by R^{10} and v^{-8}, respectively, while the quadrupole contributions to the transition cross-sections (σ_D and σ_A) and the overlap integrals are substituted. In the most general case the contributions of all interactions to the total transfer probability can be summed:

$$W_{DA} = W_{DA}(d-d) + W_{DA}(d-q) + W_{DA}(q-d) + W_{DA}(q-q) + \ldots \qquad (3)$$

A. S. Davydov analyzed the resonant interaction between atoms from different positions [36]. He considered a system of two identical atoms under conditions where one atom is in the excited state while the other is in the ground state. Using a first approximation of nonstationary perturbation theory it was possible to calculate the corrections to system energy under conditions of interaction between its constituent atoms:

$$\Delta E_1(R) = \langle \psi_1 | H | \psi_1 \rangle, \quad \Delta E_2(R) = \langle \psi_2 | H | \psi_2 \rangle. \qquad (4)$$

These corrections are determined by the matrix element of the interaction energy and in the case of two dipoles can be expressed as:

$$\Delta E_1 (R) = \Delta E_2(R) = e^2 \hbar f_{n0} \, \Phi (1, 2)/2\mu \, \omega \, R^3. \qquad (5)$$

Here f_{n0} is the oscillator strength of the electron transition between the ground and excited states of the atom; $\Sigma(1, 2)$ is the geometric factor dependent on the orientation of the dipole transitions in both atoms; μ, e is electron mass and charge; ω is electron transition frequency; and R is the distance between atoms.

Accounting for the time dependence of the wave functions ψ_1 and ψ_2 the author [36] obtained an oscillating solution where the excitation transition time from one atom to another and back, i.e., the excitation exchange time between atoms was equal to

$$\tau = \pi \mu \omega R^3/e^2 f_{n0} \, \Phi (1, 2) . \qquad (6)$$

A. S. Davydov was therefore the first to demonstrate that resonance interaction leads to splitting of the energy levels of the order of the interaction energy. Another nontrivial result is that the rate of electron excitation transfer τ^{-1} is proportional to the first power of the matrix element of interaction energy rather than the second power as in (1) and (2).

It initially seemed that the Förster-Dexter model and the Davydov model, which coexisted, were mutually exclusive and this led to discussion. Refs. [37, 38] expressed doubts as to the validity of applying first order perturbation theory to calculating the energy transfer probability in the

Förster-Dexter theory. Dexter, Förster, and Knox [39] substantiated this approach and analyzed the applicability criteria.

An analysis of the assumptions and applicability conditions of these two models reveals that in fact they are in no way contradictory but rather simply represent the two limiting cases of the more general nonradiative transfer model. Indeed, the authors of the first model formulated their applicability conditions in precisely this manner. First a certain distribution function of the final states must exist. This condition is satisfied in systems with homogeneous broadening of the donor emission and acceptor absorption spectra. The second condition which must be satisfied if we are to have an actual excitation transfer process is the requirement for a rapid relaxation process at the acceptor. This condition states that a reverse energy transfer from the acceptor to the donor D ← A is impossible, i.e., the energy transfer process in the first model is assumed to be weak compared to the relaxation processes.

The second model (the Davydov model) on the other hand generally neglects the possibility for a differential in the excitation energy before and after transfer ($h\nu_D - h\nu_A = 0$) and also ignores any relaxation processes. This means that the author considers resonant interaction to be a stronger process compared to all other processes.

Therefore, the Förster-Dexter model is a weak interaction model, while the Davydov model is a strong interaction model.

The more general problem of energy transfer in a three-level system which includes both these limiting cases was first considered by A. I. Burshteyn [40]. His study calculated the probability of energy transfer from the donor level (level 1) to the acceptor level (level 2) by means of quasi-static (dipole-dipole) interaction and from this level to the third level — the nonresonant term — by means of impact deactivation. Three cases are identified: the first case when transitions 1-2 are purely static in nature (noncoherent energy transfer), and the second, dynamic case (coherent transfer).

Beginning in 1969 the number of theoretical studies of various authors devoted to donor-acceptor energy transfer taking resonance shift into account, as well as transverse (phase) and longitudinal relaxation, increased rapidly [41-57]. The most general approach which makes it possible to obtain a complete solution to the problem is the density matrix method. Without dwelling on the features of all studies yielding similar results we will consider the most characteristic cases of nonradiative transfer for a pair of ions D and A accounting for the population relaxation rates τ_D^{-1}, τ_A^{-1}, and the phase relaxation rate $T_2^{-1} = 2\pi(\delta_D + \delta_A)$, where δ_D and δ_A are the homogeneous transition widths, together with the energy resonance shift $\epsilon = h\nu_D - h\nu_A$. Three important cases are identified here.

72   *Alimov et al.*

Weak interaction (irreversible transfer, the Förster–Dexter model). Its applicability criterion is a small excitation transfer rate (probability) W_{DA} compared to the phase relaxation rate and the population relaxation rate at the acceptor: $W_{DA} \ll T_2^{-1}$ and $W_{DA} \ll \tau_A^{-1}$. This is the case in the majority of practical models as suggested by the fact that it is the most popular object of study. We will examine the various mechanisms behind weak interaction in greater detail below.

The primary characteristic of transfer and the one determining its efficiency is W_{DA}. The donor lifetime and the quantum efficiency of its luminescence here are equal to: $\tau = (\tau_D^{-1} + W_{DA})^{-1}$, $\eta_D = (1 + W_{DA}\tau_D)^{-1}$.

Strong incoherent interaction. The criterion for this case is: $T_2^{-1} \gg W_{DA} \gg \tau_A^{-1}$. Resonant interaction in this case is reversible, since relaxation at the acceptor can no longer rapidly extract excitation from the system: $W_{DA} \gg \tau_A^{-1}$. However, fast phase relaxation $(T_2^{-1} = 2\pi(\delta_D + \delta_A) \gg W_{DA})$ makes it possible to use the concept of the probability W_{DA} when calculating the kinetics of the transfer process. In this situation there is rapid equalization of the populations of the excited levels of the donor and acceptor, $n_D^*/n_D = n_A^*/n_A$, after the instantaneous excitation of the donor over a time of W_{DA}^{-1}. The decay of both the donor and the acceptor levels then proceeds exponentially with the constant $(1/2\tau_D + 1/2\tau_A)$. It is clear that the interaction value W_{DA} characterizes only the rate of equalization of the populations and has no effect on the further kinetics of ion decay.

The quantum efficiency of luminescence of the donors will also cease to depend on W_{DA} for $W_{DA} \gg \tau_A^{-1}$ and will be determined solely by the relaxation rates of the donor and acceptor: $\eta_D = \tau_D^{-1}/(\tau_D^- + \tau_A^-)$.

From the viewpoint of energy transfer from the donor to the acceptor strong interaction will result in saturation of the transfer process or, as it is occasionally called, saturation of run-off at the acceptor.

The strong and weak interaction cases can be described using the same approach [40, 43] by characterizing the quenching rate with the generalized probability $U = W_{DA}/(1 + W_{DA}\tau_A)$, which coincides with W_{DA} only in the case of weak interaction $(W_{DA}\tau_A \gg 1)$ and is determined by the deactivation rate of the acceptor $1/\tau_A(\tau_D = 0)$ in the case of strong interaction $(W_{DA}\tau_A \gg 1)$.

A rise in the interaction probability W_{DA} with small W_{DA} therefore causes a proportional growth in quenching $U \sim W_{DA}$; in the range of large W_{DA} transfer is saturated due to the fact that acceptor relaxation becomes the "narrow throat" of the process and $U = $ const.

Strong coherent interaction. This case occurs when the matrix element of the interaction energy $\langle |H_{DA}| \rangle$ exceeds all relaxation rates in the system: $\langle |H_{DA}| \rangle \gg T_2^{-1}, \tau_D^{-1}, \tau_A^{-1}$.

As noted above, the Davydov model belongs to this case. Taking into account the imprecise transition resonance as well as relaxation processes yields the following expression for the diagonal elements of the density matrix (populations) [54]:

$$\rho_D(t) = \exp\left[-\left(\tfrac{1}{\tau_D} + \tfrac{1}{\tau_A}\right)\right]\left|\cos\left(\tfrac{\Omega_{2,1}}{2}t\right) + i\,\frac{\widetilde{E}_A - \widetilde{E}_D}{\Omega_{2,1}}\,\sin\left(\tfrac{\Omega_{2,1}}{2}t\right)\right|^2,$$

$$\Omega_{2,1} = \sqrt{(\widetilde{E}_A - \widetilde{E}_D)^2 + 4 <| H_{DA} |>^2},$$

$$\widetilde{E}_A = E_A - i/2\tau_A, \quad \widetilde{E}_D = E_D - i/2\tau_D.$$

Quantum oscillations in the populations of the excited levels of the donor and acceptor are observed here at times $t < T_2$ when no changes have yet occurred in excitation number nor in the phase relations. The coherent excitation exchange time between atoms (the oscillation period) remains of the order of $\tau = \Omega_{1,2}^{-1} \sim \langle|H_{DA}|\rangle^{-1}$, as in the Davydov model.

The decay rate of the oscillations is determined by the phase relaxation rate (here for simplicity we have assumed that $T_2 = \tau_A$). The lack of precise resonance of transitions D and A ($\epsilon = E_D - E_A \neq 0$) is manifested as a diminishing level of modulation of the populations, which may result in complete aperiodicity of the process when $\epsilon \gg \langle|H_{DA}|\rangle$ due to the increase of the second term under the modulus sign in Equation (7) similar to phase relaxation.

In order to characterize the final result of energy transfer (quenching, sensitization) the authors of Refs. [40, 46] proposed approximating the relation $\rho_D(t)$ by the exponent e^{-Ut} with the factor $U = [\int_0^\infty \rho_D(t)dt]^{-1}$. Here U is the generalized probability characterizing actual energy transfer. It is clear that it in no way is equal to the exchange rate τ^{-1} (see (6)).

We note that the transition energy splitting obtained by Davydov for his model remains valid and is characteristic specifically of the strong coherent interaction case.

Below we examine in somewhat greater detail the most common and important case of energy transfer in a pair of weakly interacting particles by focusing primary attention on a classification of transfer processes as resonant and nonresonant depending on the effect of the dissipative subsystem (crystal oscillations) to which the given pair is strongly coupled.

In their pioneering studies Förster [34] and Dexter [35, 58] defined the entire range of nonradiative energy transfer processes as resonant processes, referring to the actual overlap integral of the luminescence spectra of the excited ions and the absorption spectrum of the unexcited ions. Expressing the interaction probability W through the latter the authors formally avoided the issue of the nature of the optical transitions, the broadening mechanisms, and the characteristics of the observed resonance. In spite of the common nature and genuine simplicity of formally establishing a relation between the transfer probability and the spectral overlap integral such an approach is limited due to the impossibility of predicting typical behavior or any characteristic features of the various energy transfer processes prior to their detailed investigation.

The proposal of the authors of Ref. [59] to treat all transfer processes in solids as nonresonant processes since they can occur only by phonon participation represents an opposing viewpoint. Indeed, without "phonon" line broadening, resonant interaction of the two optical centers would be possible only when their electron transition frequencies were identical (in the range 10^3 - 10^5 cm^{-1}) accurate to the natural broadening level δ given by the spontaneous lifetime of the metastable level τ_0: $\delta = \hbar/2\tau_0 < 10^{-4}$ cm^{-1} for $\tau_0 = 10^{-6}$ - 10^{-2} sec. Such exact resonances can evidently be assumed to exist only in cooled or rarefied gases and atomic beams. The effect of fluctuations in density, the ligand environment, stresses and dislocations, micro- and macro-defects even on like ions in condensed matter produce substantial resonance shift ϵ manifested as strong inhomogeneous broadening of the electron transitions: $\Delta = 1.0$-10^2 cm^{-1}. Incorporation of electron-phonon interaction which is manifested as spectral broadening and extended low- and high-frequency phonon wings [60-65] can compensate both comparatively minor (1.0-10 cm^{-1}) and major (10-10^3 cm^{-1}) resonance shift, thereby making possible the interaction of like and unlike optical centers in solids.

The authors of Ref. [59] propose accounting for both ion-ion and electron-phonon interaction simultaneously for both interacting centers in calculating the energy transfer mechanisms. While recognizing the uniformity and thoroughness of such an analysis of the interaction processes in solids it should be pointed out that for most cases such an approach substantially complicates the understanding of the energy transfer process itself as well as the calculation of collective transfer models. For example, in the case of Raman "phonon" spectral broadening such an analysis would require simultaneous incorporation of different combinations of two-center and four-phonon energy transfer processes, which severely complicates the picture.

We believe it is more useful to classify energy transfer processes in two groups: resonant and nonresonant processes. The authors of Refs. [14, 66] have proposed classifying as resonant processes those in which there is an overlap of only the zero-phonon lines of the donor luminescent and acceptor absorption spectra corresponding to the transitions without changes in the vibrational quantum number. In this classification nonresonant mechanisms are all remaining mechanisms where resonance results from an overlap of either the zero-phonon line of a single center with the phonon wing of another center or of two phonon wings, when the vibrational states of the system change during the transfer process.

The latter classification of energy transfer processes makes it possible to separate the strong phonon broadening processes of the optical transitions from the weaker energy transfer processes (the incoherent interaction model) and to investigate these separately by tracing the mutual link. Dividing all transfer mechanisms into two groups makes it possible, without knowing the specific type of electron (zero-phonon line) or vibronic (phonon wing) spectra to predict in advance the possible frequency and temperature behavior of the transfer rate in these strong optical models and to emphasize in the nonresonant transition case the direct participation of the third (or more) particle — the phonon — in the interaction between the two electron centers.

We will use this classification below to analyze the elementary interaction processes for two optical centers in a solid. The common assumption to all mechanisms is that the phonons in the process have a wavelength less than of the order of the ion-ion distance which makes it possible to neglect interference effects in the calculations [59] which substantially reduce interaction efficiency.

1.1. Nonresonant Energy Transfer

Here we are discussing interaction models in which energy transfer from one optical center to another requires simultaneous creation or destruction of at least a single phonon [59, 67-72].

One-phonon energy transfer. The case of energy transfer involving a single phonon whose energy compensates the resonance shift of the electron transitions of the interacting centers was examined for the first time in Refs. [67, 68, 71]. The schemes of such a process for the case of a Stokes transition with phonon emission (probability $W^{\leftarrow}$) and for anti-Stokes transfer with phonon absorption (probability $W^{\rightarrow}$) are shown in Figure 1a and b.

For the Debye model of the phonon state density the interaction probability can be given as [59]

$$W^{\rightleftarrows} = \frac{J^2 (f - p)^2}{\pi \hbar^4 \rho} \left(\sum_s \frac{a_s}{v_s^5} \right) \left\{ \begin{array}{c} n(\epsilon) + 1 \\ n(\epsilon) \end{array} \right\} \epsilon. \tag{8}$$

Here $J = \langle \varphi_I' \varphi_{II} | H_{I\,II} | \varphi_I \varphi_{II}' \rangle$ is the matrix element of the interaction of ions I and II; f and p represent the electron-phonon coupling strength for the ground and excited levels, respectively; ρ is the density of the medium; a_s and v_s is a constant of the order of unity and the speed of sound in the medium for mode s, respectively.

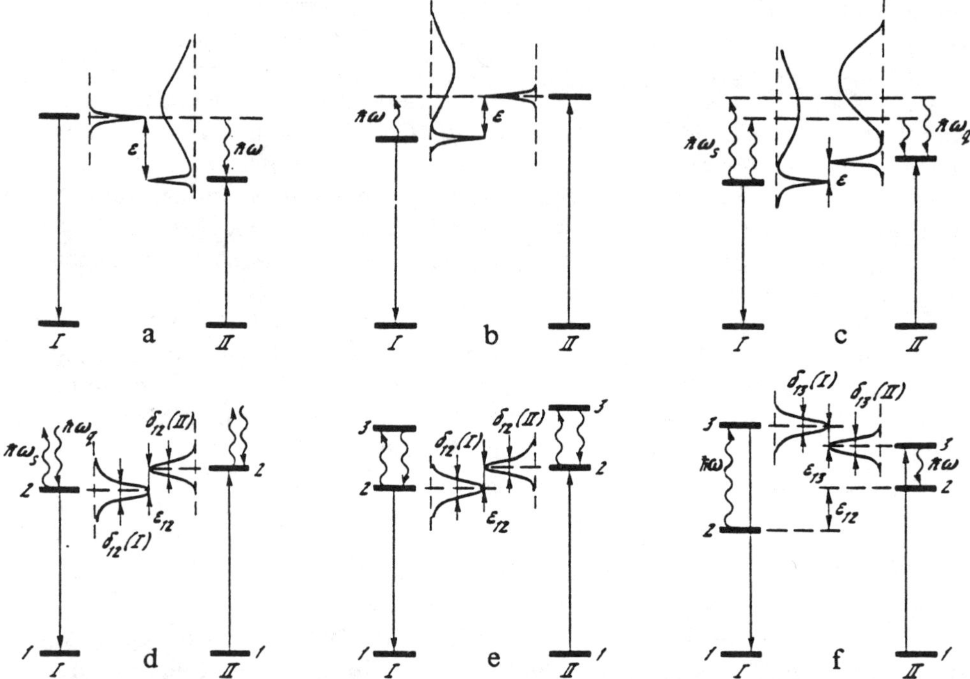

Fig. 1. Elementary nonresonant (a-c) and resonant (d-f) interaction schemes for the pair of particles 1 and 2.

A characteristic feature of this type of process is its proportionality to the frequency shift ϵ. Selection of the upper or lower term in braces corresponds to a process with phonon emission ($W^{\rightarrow}$) or phonon absorption ($W^{\leftarrow}$) where

$$n(\epsilon) = [\exp(\epsilon/kT) - 1]^{-1} \tag{9}$$

is the Bose-factor characterizing the population of the phonon sublevels. The quantity ϵ is equal to resonance shift of the electron transitions of interacting ions ($\epsilon = |E_{\mathrm{I}} - E_{\mathrm{II}}|$) and corresponding to the energy of the phonon involved in the process. It is clear from Expression (8) that the relation of the probabilities of the transfer processes involving phonon emission $W^{\rightarrow}$ and absorptions $W^{\leftarrow}$ is determined solely by the occupation number of the phonon states $n(\epsilon)$:

$$W^{\leftarrow} = W^{\rightarrow} e^{-\epsilon/kT} \tag{10}$$

Characteristically in the low-temperature limit ($kT \gg \epsilon$) the energy transfer process with phonon emission is rather efficient and temperature-independent ($n(\epsilon) \ll 1$) at the same time that energy transfer with phonon absorption with a substantially lower probability has a sharp exponential dependence on T (see Expression (10)). In the high temperature limit ($kT \gg \epsilon$) the probabilities of both processes are equal (see (10)) and become linearly-dependent on temperature ($n(\epsilon) = KT/\epsilon \gg 1$) while they are independent of the phonon energy and the resonance shift frequency.

We note that the one-phonon nonresonant energy transfer mechanism examined above can be treated within the framework of the overlap integral between the zero-phonon line and the Stokes $\overrightarrow{I_V}$ (or anti-Stokes $\overleftarrow{I_V}$) vibronic wing. Here the intensity of the vibronic spectrum is described by an expression similar to Expression (8):

$$I_V^{=}(\epsilon) \sim (f-p)^2 \epsilon \begin{Bmatrix} n(\epsilon)+1 \\ n(\epsilon) \end{Bmatrix}, \tag{11}$$

where the variable ϵ is measured from the zero-phonon line and corresponds to the frequency of the phonon in the vibronic transition.

A comparison of Expressions (8) and (11) emphasizes the lack of a dependence of the energy transfer on the width of the zero-phonon line δ and, therefore, on its temperature dependence $\delta(T)$. This important feature of the energy transfer mechanism will become obvious in the calculation of the overlap integral of a narrow zero-phonon line with the broad vibronic wing.

Ref. [64] has indicated the importance of taking into account a different formation mechanism of the one-phonon vibronic spectrum (as proposed by Van Vleck) for rare-earth ions.

Unlike the Franck-Condon mechanism (see Expression (11)) which is determined by the diagonal matrix element of the electron part of the vibronic interaction (VI) operator and which is dominant for the allowed optical transitions the Van Vleck mechanism accounts for the nondiagonal vibronic interaction element and may become determinant for parity-forbidden optical transitions of rare-earth ions which are allowed due to the mixing of states of opposite parity by the odd dynamic crystal field component.

The latter mechanism results in a stronger dependence of the one-phonon transfer probability W or the vibronic wing intensity I_V on the gap ϵ or energy of the phonon $\hbar\omega_{ph} = \epsilon$ involved in the process [64].

For the case of noncentrosymmetric ion positions

$$W_{nc}^{=} \sim I_V^{=} \sim \epsilon^3 \begin{Bmatrix} n(\epsilon)+1 \\ n(\epsilon) \end{Bmatrix}, \tag{12}$$

while for the centrosymmetric positions

$$W_{\text{nc}}^{=} \sim I_{\text{V}}^{=} \sim \epsilon^5 \left\{ \begin{array}{l} n(\epsilon) + 1 \\ n(\epsilon) \end{array} \right\}$$

It is important here that the dependence on ϵ does not vanish even at high temperatures ($kT \gg \epsilon$), when $W_{\text{nc}}^{=} \sim \epsilon^2 kT$, while $W_{\text{cc}} \sim \epsilon^4 kT$.

Two-phonon energy transfer (diagonal vibronic interaction). The scheme of this process is shown in Figure 1c. In this energy transfer model the interaction process involves absorption of a phonon of a single frequency $\hbar\omega_s$ and creation of a phonon of a different frequency $\hbar\omega_q$ with compensation of frequency shift of the electron transitions: $|\hbar\omega_s - \hbar\omega_q| = \epsilon$. Characteristically, phonons are created and destroyed in isolation at centers 1 and 2 involved in the interaction.

In accordance with the results from Ref. [59] the expression for the energy transfer probability in such a model takes the form

$$W = \frac{J^2 (f-g)^4}{2\pi\hbar^7 \rho^2} \left(\sum_s \frac{a_s}{v_s^5} \right) (kT)^3 I_2 . \tag{14}$$

Here the function

$$I_2 = \int_0^{\Theta/T} \frac{x^2 dx}{(e^x - 1)(1 - e^{-x})}, \quad x = \frac{\hbar\omega_{s,\,q}}{kT} , \tag{15}$$

accounts for the contribution of all possible oscillations $\hbar\omega$ corresponding to resonance $\hbar\omega_s - \hbar\omega_q = \epsilon$ and involved in the nonresonant interaction mechanism. The result given here was also obtained in a Debye approximation of the phonon spectrum when ϵ is much less than $\hbar\omega_s$ and $\hbar\omega_q$.

The authors of Ref. [59] consider the dependence on shift frequency as well as the cubic nature of the temperature dependence $W \sim T^3$ at temperatures below the Debye temperature $\Theta (I_2 \simeq 2)$ and the square-law nature of $W \simeq T^2$ for $T > \Theta (I_2 \simeq \Theta/T)$ to be a characteristic feature of this energy transfer model.

The authors of Ref. [59] treated this energy transfer mechanism as the only mechanism that cannot be described within the framework of the overlap integral of the donor emission and acceptor absorption spectra. However we cannot agree with this position since this transfer model is easily understood in the language of the resonance of the anti-Stokes vibronic luminescence wing of center I and the Stokes vibronic absorption wing of center II.

We can represent the desired probability as the overlap integral of the vibronic spectra noted above (expression (11)):

$$W \sim \int I \overleftarrow{\,}_{\text{V,I}} \overrightarrow{\Gamma}_{\text{V,II}} d\omega \sim (f-p)_{\text{I}}^2 (f-p)_{\text{II}}^2 \int_0^{k\Theta} (\hbar\omega + \epsilon) n(\hbar\omega + \epsilon) \hbar\omega \, [n(\hbar\omega) + 1] \, d\omega. \tag{16}$$

Here as in Ref. [59] we can assume that the electron-phonon coupling energy is identical for both interacting optical centers: $(f - p)_I = (f - p)_{II}$.

After substitution of the expression for $h(\hbar\omega)$ and substitution of the variable $\hbar\omega$ by $\hbar\omega/kT$ under the integral sign we obtain an expression similar to Expression (14) for $\epsilon \ll \hbar\omega$.

1.2. Resonant Energy Transfer

We examine another group of interaction models which, in accordance with our classification, will be referred to as a resonant group, although we recall that at least two phonons are required for its realization in solids.

Resonant transfer involving Raman scattering of phonons. The scheme of this process is shown in Figure 1d. In the first stage of this process phonon scattering occurs at one or both of the optical centers (instantaneous absorption of a phonon of one frequency $\hbar\omega_s$ and emission of a phonon of a different frequency $\hbar\omega_q$), which results in relaxation of the phase memory, spectral broadening of the electron transition and compensation of resonance shift of the interacting centers. In the second stage energy is exchanged between sites I and II. The expression for the probability of this process in accordance with the results of Ref. [59] takes the form

$$W = \frac{J^2 \, (f'' - p'')^2}{4\pi\hbar^7 \rho^2 \epsilon^2} \left(\sum_s \frac{a_s}{v_s^5} \right)^2 (kT)^7 I_6 , \tag{17}$$

where

$$I_6 = \int_0^{\Theta/T} \frac{x^6 dx}{(e^x - 1)(1 - e^{-x})} \; ; \; x = \frac{\hbar\omega}{kT} \; ; \tag{18}$$

f'', p'' are the second order electron-phonon coupling coefficients. This result was obtained assuming that ϵ is much less than kT and $\hbar\omega$. The expression becomes more complex when these conditions are not satisfied.

It follows from Expression (18) that phonons of energy $\hbar\omega \simeq 6 \, kT$ make the primary contribution to I_6 and therefore to the transfer process itself. This permits calculation of I_6 either by direct substitution of $x = 6$ into I_6 ($I_6 = 678$) or by using the approximation $I_n = n!$ ($I_6 = 720$) [59]. This routine which is responsible for the high power of the relation $W(T) \sim T^7$ is valid only for phonon frequencies $\hbar\omega$ below the Debye frequency ($\hbar\omega < k\Theta$) and for temperatures $T < \Theta/7$. Substitution of the approximation $I_6 \simeq (1/5)(\Theta/T)^5$ into Expression (17) results in a square-law variation in the transfer rate as a function of temperature — $W \sim T^2$ — for the very high temperature range ($T > \Theta$). In the intermediate temperature range $\Theta/7 < T < \Theta$ the law varies continuously from T^7 to T^2 which requires using an exact expression for I_6. An expression analogous to (17) is valid for the Kramers ions with an odd number of electrons and a half-integer spin. However, this relation has a stronger temperature dependence: $W \sim \epsilon^{-2}(kT)^9 I_8$.

A characteristic feature of this energy transfer model is the precipitous drop in the probability of the process with increasing resonance shift ($W \sim \epsilon^{-2}$) which differentiates this scheme from the models examined previously.

Resonant interactions involving direct vibrational transitions between the lower electron levels. So far we have analyzed a resonant interaction model in which two electron levels (1 and 2 of each particle) are involved in the interaction. Below we consider a number of resonant schemes that depend on a third electron level (level 3) of one or both interacting ions. It is most important to account for the new electron state when it is nearby the ground or metastable states. This situation is commonplace in rare-earth ions with a Stark splitting of the electron levels in solids amounting to several tens or hundreds of cm^{-1}. Schemes of possible processes of this type are shown in Figure 1, e, f. Here the interaction of optical centers with phonons resulting in spectral broadening and compensation of resonance shift appears as direct nonradiative transitions between the lower-lying electron sublevels.

According to Ref. [59] the expression for the probability of such processes (see Figure 1e, f) can be given as

$$W = W^\partial + W^e,\tag{19}$$

where

$$W^\partial = \frac{J_{12}^2\,|D|^2\,(\Delta E)^3}{4\pi\hbar^3\rho\epsilon_{12}^2}\left(\sum_s \frac{a_s}{v_s^5}\right) n\,(\Delta E);\tag{20}$$

$$W^e = \frac{J_{13}^2\,|D|^2\,(\Delta E)^3}{2\pi\hbar^3\rho}\,e^{-\Delta E/kT}\left(\sum_s \frac{a_s}{v_s^5}\right)\frac{n\,(\Delta E)+1}{\epsilon_{13}^2 + (2\sqrt{2}\delta_{13})^2}\tag{21}$$

Here ΔE is the energy distance between the second and third proximate levels (Stark or spin-orbit splitting for the rare-earth ions), while $|D|$ characterizes the vibrational coupling strength between them, thereby determining the matrix element of the one-phonon transition:

$$\langle\psi_2\,|H_\Phi|\,\psi_3\rangle = D\xi,$$

where ξ is the stress operator. The difference between the scheme in Figure 1f and that in Figure 1d is the elimination of the electron transition involved in the ion-ion interaction process (J_{13} is used in place of J_{12}), which may, with different oscillator strengths of these transitions, result in a substantial difference in the interaction rates. The Boltzmann multiplier $\exp(-\Delta E/kT)$ in Expression (21) determines the equilibrium population of level 3 relative to level 2 which is required to achieve transfer in accordance with the scheme in Figure 1f. The lack of correlation between energy shifts ϵ_{12} and ϵ_{13} and the substantial homogeneous width ($\delta_{13} > \delta_{12}$) established by the rate of the nonradiative transition $3 \to 2$ (δ_{13} unlike γ_{12} does not vanish even as $T \to 0$), may cause there to be a lack of dependence of the transfer rate W on shift ϵ_{12}.

As we have noted above resonant transfer processes and their dependences on frequency shift and temperature can be easily and clearly described in the overlap integral formalism relating to the overlap of the homogeneously-broadened zero-phonon lines (see Refs. [34, 35]). Taking into account that the overlap integral of the two Lorentz contours (homogeneous contours) is also Lorentzian [12] we can write the following for the transfer probability:

$$W = W^{\partial} + W^{e} \sim J_{12}^{2} \frac{1}{1 - e^{-\Delta E/kT}} \frac{\delta_{12}(I) + \delta_{12}(II)}{\epsilon_{12}^{2} + [\delta_{12}(I) + \delta_{12}(II)]^{2}} + \qquad (22)$$

$$+ J_{13}^{2} \frac{e^{-\Delta E/kT}}{1 - e^{-\Delta E/kT}} \frac{\delta_{13}(I) + \delta_{13}(II)}{\epsilon_{13}^{2} + [\delta_{13}(I) + \delta_{13}(II)]^{2}} .$$

Here the second cofactor after the squared matrix interaction element (J_{12}^{2} or J_{13}^{2}) in both terms represents the Boltzmann population of the second or third level, respectively. The third Lorentzian cofactor is the overlap integral of the two Lorentz lines of center I and II at the $1 \to 2$ (δ_{12}) and $1 \to 3$ (δ_{13}) transitions having resonance shifts of $\epsilon_{12} = E_{12}(I) - E_{12}(II)$ and $\epsilon_{13} = E_{13}(I) - E_{13}(II)$. We note that Expression (22) is common to the resonant energy transfer processes involving both Raman phonon scattering and when the vibrational transitions between levels are taken into account.

Expressions (17), (20), and (21) can be obtained from (22) in an approximation of moderately high temperatures ($kT \ll \Delta E$) and low-level homogeneous broadening ($\delta_{12} \ll \epsilon_{12}$). It is only necessary to substitute values corresponding to any spectral broadening mechanism of the electron transition into Expression (22) for this purpose [63, 65, 73-77].

For example, with Raman broadening for non-Kramers ions with an even number of electrons

$$\delta(T) \sim I_{6}(kT)^{7}, \qquad (23a)$$

for Kramers ions with an odd number of electrons

$$\delta(T \sim I_{8}(kT)^{9}. \qquad (23b)$$

Under conditions of one-phonon relaxational broadening [59]

$$\delta(T) = \frac{|D^2|(\Delta E)^3}{4\pi\hbar^3\rho}\left(\sum_s \frac{a_s}{v_s^5}\right)\left\{\begin{array}{l} n(\Delta E)+1 \\ n(\Delta E) \end{array}\right\} \tag{24}$$

Here, as before, the upper term in braces corresponds to the 3-2 nonradiative transition with emission of a phonon of frequency $\hbar\omega = \Delta E$, thereby broadening level 3 (δ_3), while the lower term corresponds to the 2-3 transition involving absorption of a phonon $\hbar\omega = \Delta E$ and causing broadening of level 2 (δ_2). It is clear from expressions (22)-(24) that the precise establishment of the specific electron transition broadening mechanism can make it possible to reliably predict the specific resonant ion-ion interaction mechanisms involving vibrations of the crystalline lattice. It is similarly possible to take into account the specific energy transfer and broadening mechanisms determined by dynamic disorder and characteristic of only certain condensed matter [78-80].

In concluding our analysis of the elementary cases of interaction for a pair of optical centers we note that the dominance of the resonant or nonresonant mechanism in specific interaction schemes in each experimental situation will be determined by the electron-phonon interaction force, the frequency shift ϵ, and the temperature and can be determined experimentally based on the temperature and frequency dependences of the energy transfer probability.

The interaction models examined here demonstrate the fundamental parameters and conditions for spectrally-selective (sensitive to energy shift) and spectrally-nonselective (independent of ϵ) ion-ion interactions which represent the fundamental basis of the energy migration processes in a disordered set of spectrally-inhomogeneous centers examined below.

We emphasize that we have so far considered cases of energy transfer that refer to either the D-A particle pair or to a group of particles consisting of such pairs of interacting ions. When investigating actual objects — crystals and other solids — we commonly encounter more complex situations. In the more general (and more realistic) case we are dealing with two impurity subsystems that are randomly or uniformly mixed. In this case the atoms can decay both in a regular manner, thereby forming a crystalline lattice in "concentrated" crystals or randomly in "diluted" crystals. When investigating interaction and energy transfer in such a complex system consisting of a large number of different particles it is important to remember, first, that there is a substantial statistical variability of the donor and acceptor environment and, second, there is a distribution of resonance shifts ϵ of the donor-donor and donor-acceptor pairs due to variations in the local electrical field in the medium and, third, possible donor-donor and acceptor-acceptor interactions exist resulting in excitation migration within each subsystem.

2. The Kinetics of Irreversible Static Donor-Acceptor Transfer in a Particle Ensemble

Refs. [34, 81] have investigated the deactivation kinetics of the donor level by dipole-dipole energy transfer to a disordered acceptor group in the static limit when there is no motion of the donor excitations, while Refs. [82, 83] generalize the results to interactions with higher degrees of multipolarity: dipole-quadrupole ($S = 8$) and quadrupole-quadrupole ($S = 10$) interactions.

Accounting for the statistical differential of the acceptor configurations for the various donor ions results in a rather broad distribution of their deactivation probabilities which in turn is manifested as general nonexponential decay kinetics for the entire set of donor ions (see experimental studies [6, 7, 13, 15, 84-87]). These kinetics reflect the ion-ion interaction processes characterized by the efficiency microparameter C_{DA} and the interaction multipole order $S = 6, 8, 10$, as well as the spatial distribution of the ions $n(R)$.

Neglecting the radiative decay channel (over time τ_0) which is easily accounted for in the final expression and by multiplying the solution by $\exp(-t/\tau_0)$ it is possible to identify two time sections on the kinetic luminescence decay curve. The first section in the initial kinetic stage is exponential:

$$I(t) = \exp(-W_{max}t) \tag{25}$$

and is characterized by the maximum possible quenching rate for this ion set W_{max}. The macroparameter W_{max} will be determined by the sum of the elementary ion-ion interaction probabilities $W_i = C_{DA}/R_i^S$ over all possible N sites of the impurity subsystem taking account of its occupation factor or the specific concentration of active ions $c = n/n_{max}$ [19, 20]:

$$W_{max} = cC_{DA} \sum_{i=1}^{N} R_i^{-S} \tag{26}$$

We note that even in the simplest case for crystals with a high degree of order calculations using this formula are complex since they require knowledge of the complete lattice sum for each type of crystal. For disordered media (compound crystals, glasses, and liquids) the concept of the lattice sum is not generally consistent, and it is not always possible to go to integration. It has been more convenient to use the approach proposed in Refs. [27, 85, 87] in which the total lattice sum is replaced by interaction with only nearest neighbors. Calculations of specific systems [15, 85, 88] have demonstrated that interaction with the first coordination shell makes a decisive contribution to quenching efficiency due to the strong exponential dependence on distance (R^{-6}, R^{-8}, R^{-10}). Here the expression for W_{max} can be given as [87]:

$$W_{max} = cC_{DA}aR_{min}^{S},\qquad(27)$$

where a is the nearest neighborhood coordination number equal to the maximum possible number of nearest neighbors, while R_{min} is the minimum possible intervening distance between impurity ions (the cation sublattice constant for simple crystals).

Expression (27) can also be used successfully for disordered media (crystals, glasses, and liquids) if the spatial distribution of their active impurities can be modeled by a distribution of solid spheres corresponding to the impurity ion or a complex of radius $R_{min}/2$ and a coordination number of complex packing of a. The lack of exact information on the values of a and c in disordered media or in rarely-studied crystals will at first glance result in chaos in calculations using Formula (27). However, an analysis of the three types of cubic packing of impurity complexes with a = 12, 8, and 6 (face-centered, cubic-centered, and simple) and for tetragonal packing (a = 4) reveals that at a fixed impurity concentration and a constant R_{min} the product ac will be equal to $8.5\,nR_{min}^{3}$, $6.15\,nR_{min}^{3}$, $6.0\,nR_{min}^{3}$, and $6.15\,nR_{min}^{3}$; i.e., it varies little for the lattices examined above. This makes it possible to write an approximate expression for W_{max} with an accuracy sufficient for a number of experiments [87]:

$$W_{max} = 7\,C_{DA}\,nR_{min}^{3-S},\qquad(28)$$

from which it is possible to establish a relation between the macroquantities W_{max} and n measured by experiment and the microparameters C_{DA}, S, and R_{min} characterizing both the interaction and relative position of the ions.

The exponential section at the beginning of the kinetic decay curve examined here is often called the ordered section, since the donors which donate energy to the ordered acceptor set located at a fixed distance of R_{min} luminesce in this case. We note that in spite of the simplicity of determining W_{max} in the ordered section of the decay curve, Expression (28) alone cannot be used to directly find the three constituent macroparameters: C_{DA}, S, R_{min}.

A second, nonexponential, region is observed after the ordered region on the kinetic luminescence decay curve. This section reflects the decay kinetics of the donors for which the acceptors are at the minimum possible distance. Bearing in mind the random nature of acceptor distribution for $R_i > R_{min}$ it is clear that each such donor will have an individual (differing from all other) deactivation rate which will result in a sharply nonexponential luminescence decay kinetics of the donor set. Such decay (called Förster decay or disordered decay) can be represented by the expression

$$I(t) = \exp\{-\int_0^{\infty} [1 - \exp(-C_{DA} R^{-S} t)]\, n(R)\, 4\pi R^2 dR\}. \tag{29}$$

In the general case particle concentration can be a function of ion-ion distance. Ref. [87] proposed recovery of the distribution function $n(R)$ for impurity ions in matter based on the kinetics of $I(t)$ by solving the inverse problem.

It is possible to obtain a simple expression for the decay kinetics[1] [82, 83] for the case of a random particle distribution, yet a fixed density (n = const), and neglecting their actual dimensions ($R_{min} \rightarrow 0$):

$$I(t) = \exp(-\gamma_F t^{3/S}) = \exp[-(W_F t)^{3/S}], \tag{30}$$

where the macroparameter γ_F and the average rate W_F are related by the relation

$$\gamma_F = W_F^{3/S} = (4/3)\pi\, \Gamma(1-3/S)\, n C_{DA}^{3/S}. \tag{31}$$

The importance of Expression (30) lies in the fact that by analyzing this expression it is possible to directly determine two independent parameters: the multipole order S and the macroparameter γ_F or W_F. Substituting these values as well as the concentration into Expression (31) we can find the microparameter C_{DA} characterizing the efficiency of the elementary interactions. However, it is important to remember that Expressions (30) and (31) were derived assuming small acceptor concentrations (diluted solutions, $c = n/n_{max} \ll 1$) and are applicable only in the initial decay stages when the approximation $R_{min} \rightarrow 0$ is not too approximate. An analysis of Expression

---

[1] In order to simplify the analysis we will not discuss exchange ion-ion interaction with a $W = W_0 \exp(-2R/L)$ probability here, since the overlap scale of their wave functions L is quite small compared to R due to screening of the $4f$-electrons. However, as demonstrated in Refs [89, 90] the kinetic curve is nonexponential in this case as well: $I(t) = \exp[-\pi L^3/6) \times c \ln^3(W_0 t)]$.

(29) for the entire time axis ($0 < t < \infty$) taking into account the relation $n(R)$ as

$$n(R) = \begin{cases} n \text{ for } R \geq R_{min}, \\ 0 \text{ for } R < R_{min} \end{cases}$$

was carried out in Ref. [26] for the case of dipole-dipole particle interaction. It provides a clear demonstration of the general mechanisms behind the conversion of the decay kinetics from an exponential law to a Förster law although its weak point is the use of integration at distances $R \approx R_{min}$ where there is a limited and discrete number of ions.

The Förster theory of static dipole-dipole quenching was generalized to the case of significant acceptor concentrations $c \sim 1$ in Refs. [20, 91] which led to the appearance of terms of second and higher orders in acceptor concentration. So far the literature has not reflected experimental data indicating the discovery of nonlinear terms in the relation $\gamma_F = f(n)$.

Using Refs. [19, 20, 62] as our basis it is possible to find the quadratic terms in the concentration in the expression for γ_F with a random multipole order: $S = 6, 8, 10$ [87]. The expression for the decay kinetics will be analogous to (30) in this case although the microparameter γ_F will be substituted with γ_F':

$$\gamma_F' = \gamma_F[1 + c(1 - 2^{3/S-1})]. \tag{32}$$

It is clear that with small specific concentrations ($c \ll 1$) the term dependent on c can be neglected and we arrive at expression (31): $\gamma_F' = \gamma_F$. The coefficient for c grows at substantial concentrations for systems with a high multipole order S and therefore the terms nonlinear in the concentration manifest themselves more strongly.

By setting equal the exponents for the ordered and disordered decays (Expressions (25) and (30)) it is impossible to determine the boundary time of the transition from one decay to another:

$$t_1 = (\gamma_F/W_{max})^{S/(S-3)} = R_{min}^S/C_{DA}, \tag{33}$$

which was equal to the inverse of the transition rate at the nearby donor-acceptor pair [87].

Expression (33) was obtained at moderately high acceptor concentrations c and in this range coincides with the more general expression obtained in Ref. [20] for a random c. It is important for t_1 to be independent of concentration over a broad range of c and to grow only with an increasing microparameter R_{min} and a diminishing decay efficiency C_{DA}. This suggests that the ordered section of the decay kinetic curve for specimens with a different yet moderately high acceptor concentration will be observed over the same time interval. However, the extent of the change in luminescence intensity and therefore the number of centers luminescing in this region will be substantially different and will depend on the total particle concentration c.

This makes it possible to relate the level of variation in intensity on the exponential section $0 < t < t_1$ to the degree of order of active ions in a disordered medium by writing the fraction of ions (donors) having a spatially-ordered set of nearest neighbors (acceptors) as[2] [87]

$$P_o = 1 - I(t_1). \tag{34}$$

P_0 can be estimated by the formula

$$P_o = 1 - \exp(-W_{max}t_1), \tag{35}$$

where $W_{max}\, t_1 = ac = 7\, nR_{min}^3$ is equal to the average number of nearest neighbors and is determined by the specific impurity ion concentration in the medium, which grows with increasing n and R_{min}. The fraction of ions having a disordered set of nearest neighbors can be given as

$$P_d = 1 - P_o = I(t_1). \tag{36}$$

We note that the two-stage nonexponential decay law examined above applies to the case of instantaneous donor excitation by a short pulse, i.e., it is assumed that the donors are uniformly excited regardless of their decay rates at initial time $t = 0$: $n_i^* \sim n_i$. Another situation occurs in the case of rapid switching-off of stationary excitation or in the case of excitation by square-wave pulses of duration of the order of or greater than the characteristic excitation transfer times: $t_p \gtrsim W_F^{-1}$. Such a case was examined in Ref. [92].

It turned out that in this case, unlike the first case (Expression (30)) the kinetics are near-exponential with a characteristic time equal to the average decay time: $\tau_{av} = \int_0^\infty I(t)t\,dt / \int_0^\infty I(t)\,dt$. Such a significant difference in the kinetics of the decay processes is due to the fact that in the stationary case a certain new distribution that differs from the equal-probability distribution arises in the period preceding the removal of excitation. As we can easily determine from the balance equations the number of excited donors of short lifetime τ_i (with a high decay probability: $\tau_i^{-1} = \tau_0^{-1} + W_i$) will be less,

---

[2] Ref. [27] employed such a treatment of the kinetics to interpret the complex Förster decay in a model of a black body (sphere) of a radius that varies over time.

while the number of donors with a large τ_i will be greater, i.e., $n_i^* \sim n_i\tau_i$. This is a fundamentally important fact for experimenters who use the luminescence decay kinetics as a tool to investigate the transfer and interaction microprocesses in activated media. It is important that the first technique utilizing instantaneous excitation is the more exact and fruitful technique, while the second method involving rapid switching-off of stationary excitation can be used only as a supplement to the first method.

Returning to Expression (30) we note that the decay rate of the excited state for the ion set is not constant for the Förster section of the kinetic curve and is equal to $\gamma_F/t^{(S-3)/S}$. If its growth is, severely limited to the maximum rate W_{max} determined by Expression (27) in the range of small t, in the range of large t it will monotonically vanish. This, however, does not suggest that the excitation lifetime $\tau \to \infty$ as $t \to \infty$. The lower limit for the decay rate of the excited state is normally limited by either spontaneous emission processes at $A = 1/\tau_0$ which we neglected in Expressions (25) and (30) or the energy migration processes among the metastable levels of the donors to the acceptors with a probability $\overline{W}(C_{DD})$ which enhances the quenching of the active ions for which the direct quenching interaction probability is small [6, 7, 13-17, 23, 85, 93, 94]: $W_i = C_{DA}R^{-S} \to 0$. It follows that the Förster nonstationary section of the decay kinetic curve demonstrating the disorder in particle configuration is within the following rate range

$$(A + \overline{W}) \leqslant \gamma_F/t^{(S-3)/S} < W_{max}. \tag{37}$$

In order to reliably determine the multipole order $S = 6, 8, 10$, from the decay kinetics with a random particle configuration, to say nothing of the more delicate problem of recovering the function $n(R)$, it is necessary to achieve as broad a range of $\gamma_F/t^{(S-3)/S}$ as possible, at least within an order of magnitude: $W_{max}/(A + \overline{W}) \geqslant 10$.

In the analysis above we considered the static quenching interaction kinetics detected experimentally in three cases:

1) in an ensemble of donor and acceptor particles of different types when there is a very low donor concentration ($n_D \ll n_A'$) the high average donor-donor distances ($\overline{R}_{DD} \approx n_D^{-1/3} \gg \overline{R}_{DA} \approx n_a^{-1/3}$) makes it impossible for migration interactions whose rate $W_{DD} = C_{DD}/R_{DD}^S$ to compete with quenching interactions: $W_{DD} \ll W_{DA} C_{DA}/R_{DA}^S$;

2) in an ensemble of like particles with two types of interaction of which one, the quenching interaction (cross-relaxation interaction), causes an irreversible decay of electron excitation (from the metastable level to the nonradiative level $W_{DA} = C_{DA}R^{-S}$) with an efficiency much greater than the other, migrational, interaction, which causes delocalization of excitation in

the space ($W_{DD} = C_{DD}R^{-S}$) yet is not related to excitation drift from the metastable level: $C_{DA} \gg C_{DD}$;

3) an irreversible nature of the migration over the metastable levels and running downward on the frequency scale is implemented due to the static spread of the metastable energies (inhomogeneous broadening) with a half-width δ at low-temperatures ($kT \ll \Delta$) in an ensemble of like particles with no cross-relaxational quenching. In this case donors with a transition frequency below the excitation frequency function as acceptors and can be relabeled: $C_{DD}(v_e \rightarrow v) = C_{DA}(v_D \rightarrow v_A)$, where $v < v_e$, while the excess energy $\epsilon = hv_D - hv_A$ is contributed to the medium as acoustic vibrations $hv_{ph} = \epsilon$.

3. Reversible Electron Excitation Migration in an Ensemble of Like Particles Without Acceptors

We know that in the absence of quenchers (acceptors) excitation energy migration will only result in a spatial delocalization of excitation: an excitation that has arisen at a certain initial center is transferred to neighboring centers until it is distributed throughout the entire system. Here the further the center the greater the probability that it will be excited by the initial center after it reaches centers in relatively close proximity.

This process can be characterized by means of an expression describing the temporal behavior of the population of the initially-excited centers $I(E_L, t)$ through the interaction macro- and micro-parameters. Given its complexity this problem was solved with certain assumptions and approximations which were then eliminated as the theory was refined and experimental data were accumulated. In spite of some progress in recent years the issue of the kinetics of $I(E_L, t)$ and its proper relation to the microscopic interaction parameters across the entire time scale remains open.

In the general case in order to consider the process of interest to us it is necessary to solve a system of coupled kinetic equations describing migration in a balance approximation [1]:

$$dP_i(t)/dt = - \sum_j [W_{ij} P_i(t) - W_{ji} P_j(t)] \tag{38}$$

with the initial condition $P_i(t = 0) = \delta_{i0}$, where $P_i(t)$ is the probability of detecting excitation at the ith center at time t; δ_{i0} is the delta-function; W_{ij} is the probability of excitation transfer from the ith center to the jth center. In the general case $W_{ji} \neq W_{ij}$.

Such an equation describes excitation delocalization from the initially-excited center to all remaining initially unexcited centers and solves exactly the problem of when the donors form a proper lattice ($c = 1$). Here the decay kinetics of the initial excitations (centers of energy E_L) for the short-range potential (SRP) of interaction can be described in the following manner [95]:

$$I(E_L, t) = P_0(t) = \exp\left(-\sum_i W_{0i}\, t\right) J_0\, (2\, t)^3, \tag{39}$$

where summation is carried out over all sites of the crystalline lattice, while J_0 is a modified zeroeth order Bessel function.

If this expression over rather short times yields an exponential decay law

$$I(E_L, t) = \exp(-t/\tau) \tag{40}$$

with a rate $1/\tau = \sum_i W_{0i}$, over rather long times expression (39) reduces to diffusion kinetics

$$I(E_L, t) = W_{\text{diff}}t)^{-3/2} \tag{41}$$

at the rate

$$W_{\text{diff}} = 4\, \pi\, n^{2/3}\, D_o. \tag{42}$$

Here D_o is the coefficient of diffusion throughout an ordered lattice:

$$D_o = (1/6) \sum_i W_{0i} R_{0i}^2. \tag{43}$$

It is clear that in this situation calculation of the diffusion coefficient is reduced to finding the probability W_{0i} and carrying out summation.

Unlike the regular lattice case examined here kinetic equation (38) cannot be solved exactly for disordered (diluted) media which include popular laser materials based on rare-earth ions (TR^{3+}). The need to average over random center configurations which cannot be done in the general case substantially complicates the problem.

Ref. [96] proposed treating the spectral-spatial electron excitation energy migration process to take into account the chaotic distribution of impurities within the framework of Förster disordered energy transfer by analyzing the experimental data of Refs. [97–102] on the relaxation of the spectral-nonequilibrium hole in the inhomogeneously-broadened luminescence line.[3]

Ref. [96] formulated an approach to the problem and proposed a simplified analysis that neglects the reverse energy transfer effects to the initially-excited centers (vanishing of the second term in Expression (38): W_{ji} 0). It has been noted that this assumption must be valid at the early stages of the spectral migration process when the population of the initially-excited donor centers substantially exceeds that of the centers on the inhomogeneously-broadened pedestal which have acquired energy by nonradiative transfer. Here the expressions for the relaxation kinetics of a spectrally-narrow packet of excited centers of energy E_L (in selective excitation spectroscopy) and analogously for a spectrally-narrow packet of unexcited centers ("de-excitation" in the spectroscopy of selective stimulated luminescence) reduces to the static quenching kinetics (see Expressions (30), (31)).

Such an approach was subsequently used on several occasions to analyze experimental data on spectral energy migration [8, 106-110] although the range of its applicability has not been precisely defined or substantiated.

In the case where the reversibility of the ion-ion transfer and the disordered distribution of active ions (diluted crystal or glass) are fully accounted for, it has been extraordinarily difficult to obtain an exact solution for the entire time scale and therefore the authors of Refs. [111, 112] have proposed an approximation that is suitable for moderately large times ($0 < t < t_2$) and accounts for the reverse transfer process only for a nearest neighbor pair for which energy exchange is the simplest. Formally this involves adding to the known Golubov-Konobeev solution for static quenching [19] (which, unfortunately, is not cited by the authors of Refs. [111, 112]) the function $f(W_{0i}t)$ which accounts for pair back-transfer, after which it appears as

$$I(E_L,\, t) = \exp\sum_i \ln \{1 + c\,[e^{-W_{0i}t}f(W_{0i}t) - 1]\} \,, \tag{44}$$

where

$$f(W_{0i}t) = 1 \tag{45}$$

for irreversible Förster transfer (model 1)

$$f(W_{0i}t) = \mathrm{ch}(W_{0i}t) \tag{46}$$

---

[3] The hole burning and relaxation effects in an inhomogeneously-broadened line have been investigated in the RF range in EPR research [103-105].

where the fast reversible energy exchange in an isolated ion pair (model 2) is accounted for exactly, while in model 3 the function $f(W_{0i}t)$ is an approximation of Expression (46):

$$f(W_{0i}t) = 1 + 1/2(W_{0i}t)^2. \tag{47}$$

Analogous to the kinetics of irreversible luminescence quenching [15, 20, 86, 87] at moderately high concentrations ($c < 1$) two time sections clearly appear on the kinetic curve $I(E_L, t)$.

The first section in the initial kinetic stage is an exponential section. It takes the form

$$I(E_L, t) = \exp(-W_{max}t) \tag{48}$$

and is characterized by the maximum possible migration rate W_{max} for the given ion set. Here as we see from Expressions (46) and (47) with small times ($t < W_{0i}^{-1}$) the function $f(W_{0i}t)$ is equal to unity for all models, while the macroparameter W_{max}, as is the case in irreversible energy transfer, will be equal to the sum of probabilities of the elementary energy transfer events among all sites of the impurity subsystem accounting for its occupation factor [19, 20]:

$$W_{max} = cC_{DD}\sum_i R_{0i}^{-S}. \tag{49}$$

As demonstrated previously for media with a disordered structure (composite crystals, glasses) where the lattice sum is unknown, a simplified expression can be used for the analysis:

$$W_{max} = 7C_{DD}nR_{min}^{3-S}. \tag{50}$$

This establishes the relation between the experimental macroparameters W_{max} and n and the microparameters C_{DD}, S, and R_{min} characterizing the interaction and relative configuration of the active ions.

In spite of the simplicity of determining W_{max} in the initial decay region, Expression (50) by itself as well as (26)-(28) cannot be used to directly find its three constituent microparameters: C_{DD}, S, and R_{min}. The exponential section of the decay kinetic curve is often called the ordered section since it contains luminescing ions that have contributed energy to the ordered set of neighbors at a fixed distance R_{min}.

A second, nonexponential, section follows the ordered section on the kinetic luminescence decay curve. This section reflects the decay kinetics of the ions for which the nearest neighbors are at distances exceeding the minimum possible distance. Bearing in mind the random nature of ion distribution for $R_{0i} > R_{min}$ it is clear that each such donor will have an individual (unique) deactivation rate which overall results in nonexponential luminescence decay kinetics for the set.

In the case of low-concentration (diluted) media where the discreetness of the active ion distribution is lost the authors of Refs. [111, 112] going from the sum to the integral in Expression [44] obtained a simple yet very important result which can be given as

$$I(E_L, t) = \exp(-\gamma_I t^{3/S}) = \exp[-W_I t)^{3/S}, \tag{51}$$

where

$$W_I = \gamma_I^{S/3}; \ \gamma_I = \frac{4}{3} \pi n \Gamma \left(1 - \frac{3}{S} \right) XC_{DD}^{3/S}. \tag{52}$$

It is clear that the decay law of the initial excitations coincides with the expressions for the nonexponential kinetics of irreversible static or Förster energy transfer accurate to the coefficient X: $\gamma_I = X\gamma_F$.

The coefficient X is determined by the model used to incorporate reverse transfer, while for model 2 which more accurately corresponds to the case of low concentrations

$$X = 2^{3/S-1}. \tag{53}$$

This coefficient can be treated as the effective fraction of "carrier" ions (of the total concentration n) actively participating in the energy migration process ($n_p = Xn$) and resulting in an irreversible transfer of excitation from the initial center, unlike the other, "idle", fraction of ions $n_i = n(1 - X)$ which are effectively excluded from the migration process due to the return of excitation from these ions to the initial center. Here the spatial distribution of the "carrier" ions will not fully correlate with the distribution of all ions n_i since the latter will have a random distribution X_i dividing the ions into "carrier" and "idle" ions. The difference between these distributions is fundamental in substantiating the applicability of the results of the hopping theory for the case where the migration and quenching interactions in a set of like ions (such as Nd^{3+} ions in solids) are simultaneously accounted for.

The values of the coefficients X (model 2) are equal to 0.797, 0.648, and 0.616 for the following three most important interaction mechanisms, respectively: dipole-dipole interaction ($S = 6$), dipole-quadrupole interaction ($S = 8$), and quadrupole-quadrupole ($S = 10$) interaction calculated by Formula (53).

The treatment of the coefficient X proposed above makes it possible to understand its decay with increasing multipole order ($S = 6, 8, 10$) as a growth in the fraction of "idle" ions and a drop in the "carrier"-ions due to the diminishing number of neighbors within the effective range of interaction with the excited center with enhanced short-range action.

The value of both Expression (51) and Expressions (30) lies in the fact that analysis of this expression makes it possible to directly determine independent interaction parameters: the multipole order S and the macroparameter γ_I (or W_I). Substituting these values together with the concentration into Expression (52) we can find the microparameter C_{DD} characterizing the efficiency of elementary excitations. It should be recalled that Expressions (51) and (52) are applicable only at low ion concentrations ($c = n/n_{max}$) and in the later decay stages when the assumption $R_{min} \to 0$ is not too approximate.

The boundary transition time from the ordered to disordered migration stage can be determined from equality of the decay rates in the exponential and nonexponential decay stages as has been done previously for the irreversible energy transfer process:

$$W_{max} = (\gamma_I/t_1)^{S/S-3}, \tag{54}$$

$$t_1 = R_{min}^S/2C_{DD} = t_0/2. \tag{55}$$

As we see with any multipolarity t_1 is half the boundary time t_0 for irreversible energy transfer. Expressions (54) and (55) are written for moderately high active ion concentrations ($c \ll 1$) and in this limit t_1 and t_0 are independent of c and will be given solely by the pair interaction efficiency at the minimum distance C_{DD}/R_{min}^S.

We proposed a method of estimating the fraction of impurity ions having a spatially ordered set of nearest neighbors P_o and the fraction of ions in the spatially-disordered set ($P_d = 1 - P_o$) in Section 2 based on an analysis of the kinetics of irreversible energy transfer.

We can proceed analogously for the case of reversible spectral migration and write

$$P_d = 1 - P_o \approx \exp(-2W_{max}t_1). \tag{56}$$

It is possible to account for the nonlinear terms in the concentration by expanding the logarithm in Expression (44) into a series in the concentration and by limiting the analysis to the first three terms of the expansion, as done in Ref. [87] for irreversible energy transfer.

The spectral migration kinetics in this case are also of the Förster type, while the expression for γ_I' takes the form:

$$\gamma_I' = \gamma_I 2^{3/S-1}\left[1 + \frac{c}{2}\left(1 - 2^{3/S-1}\right)\right]. \tag{57}$$

We can write the following for the three possible types of ion-ion interaction multipolarities:

$$\text{for } S = 6 \quad \gamma'_I = 0.707 \ \gamma_I \, (1 + 0.146c),$$
$$\text{for } S = 8 \quad \gamma'_I = 0.648 \ \gamma_I \, (1 + 0.176c),$$
$$\text{for } S = 10 \ \gamma'_I = 0.616 \ \gamma_I \, (1 + 0.192c).$$

It is clear that with large S the contribution to quenching kinetics from energy transfer between closely-proximate ions most subject to a concentration nonlinearity grows.

The good approximation of the nonselective spectral migration process using simple models of reversible Förster energy transfer demonstrated above is quite important for the treatment of experimental results, for determining the macro- and microparameters and for analyzing the concentration and temperature dependences of donor-donor interaction. However, the approximation does not describe the latter stages in the migration process for which the partial back transfer in solely the nearest neighbor pairs is insufficient and requires a more comprehensive incorporation of all back trajectories to the initially-excited center. Ref. [20] employed a diagrammatic technique to obtain a kinetic equation describing the dynamics of the donor population averaged over different donor configurations to account for the reversible donor-donor interactions. A solution in an approximation linear in donor concentration was obtained for migration in a disordered donor set, and although the solution does not describe the long-range asymptotics of the process, it nonetheless yields the time dependence of a coefficient similar to the diffusion coefficient in the early, nonstationary migration stage: $D \sim t^{-1/6}$.

At the same time the authors of Ref. [23] noted the impossibility of directly using the diffusion equation to analyze excitation migration in a spatially disordered medium $(4/3 \ \pi n R^3_{min} < 1)$ due to the substantial fluctuations in the ion-ion transition probability. However in order to overcome the calculation difficulties associated with averaging the kinetic equations over a random ensemble of particles they proposed a simplifying technique that involved replacing the "actual" distribution of the diffusion coefficients by a certain average quasidiffusion coefficient D_{av} which determines the average expansion time of the donor excitation packet in space. In this technique the fundamental element is not the formal method but rather the physically-substantiated method of selecting D_{av} which made it possible for the authors, by determining the upper limits for integration over space, to predict the iteration exponential dependence of the quasidiffusion coefficient on concentration: $D \sim n^{4/3}$.

We considered the application of such an approach to analyzing excitation migration in a spectrally-disordered medium where the migration rate is also variable and is modulated as a function of energy shift and the ratio between homogeneous and inhomogeneous broadening in Refs. [13, 93, 94]. It was demonstrated that the most physically justified approach was to select the average diffusion coefficient or the average resonance shift $\epsilon_{av} \approx \Delta$ for the average hop probability only after averaging the migration time over the inhomogeneous set of centers and selecting the average τ_{av}. This makes it possible to avoid the drawbacks of formally averaging the diffusion coefficient over the inhomogeneously-broadened contour, which makes it impossible to account for the discreetness of the number of interacting centers [14] as well as fluctuations in the excitation rates in a spectrally-disordered medium.

One result demonstrating the importance of properly accounting for the disordered nature of diffusion in a system of spectrally-nonequivalent centers is the identification of the nonmonotonic nature of the dependence of the rate of resulting migration (delocalization) from the start site on the relation between homogeneous (δ) and inhomogeneous (Δ) spectral line broadening. Such a nonmonotonic relation with an experimented maximum in the migration rate for $\delta \approx \Delta$ was established experimentally for an entire series of yttrium-ytterbium garnet crystals [13, 93, 94].

The long-range diffusion asymptotics of the migration process in a disordered (diluted) medium was analyzed in Ref. [113] based on an integral equation derived by the author. The solution of this equation made it possible to find expressions for the spatial and spectral diffusion coefficients. The first expression takes the following form for the case of zero inhomogeneous broadening:

$$D = \frac{0,15}{2\tau_R} \, n^{-2/3}. \tag{58}$$

Publication [114], which appeared nearly simultaneously with Ref. [112], investigates migration in a set of chaotically distributed particles with the authors using a generalized (non-Markovian) theory and taking into account the quadratic terms in the expansion of the generalized diffusion coefficient in the concentration of interacting particles. As before the author neglected the effect of the acceptors, although unlike Ref. [112] they did not trace the population of the initially-excited centers but rather followed the mean square delocalization radius of the excitations in the space surrounding the initially-excited centers $\langle r^2(t) \rangle$ which is related to the spatial population profile $P(r, t)$ by the expression

$$\langle r^2(t) \rangle = \int_0^\infty P(r, t) \, r^2 \, dr. \tag{59}$$

Using scaling concepts the authors of Ref. [114] gave the functional notation for $\langle r^2(t) \rangle$ as

$$\langle r^2(t) \rangle = 3{,}966\pi n C_{DD}^{5/6} t^{5/6} [1 + 0{,}4393\pi n C_{DD}^{1/2} t^{1/2}]^{1/3}. \tag{60}$$

We note that unlike the results of Ref. [20] this expression describes not only the nonstationary-near-Förster stage but also the late (quasidiffusion) stage of the migration process.

In the early stage of the process, neglecting the second term in square brackets in Expression (60) for the excitation packet expansion kinetics, we have

$$\langle r^2(t) \rangle = 3{,}966\, \pi n C_{DD}^{5/6} t^{5/6}, \tag{61}$$

which is analogous to the relation obtained in Ref. [20]

$$D \sim \langle r^2(t) \rangle / 6t \sim t^{-1/6} \tag{62}$$

In the late stage when it is possible to neglect unity in Expression (60), $\langle r^2(t) \rangle$ grows linearly in time, which is characteristic of a diffuse migration process:

$$\langle r^2(t) \rangle = 3{,}02\pi^{4/3} n^{4/3} C_{DD} t. \tag{63}$$

Writing (63) in the form generally accepted for describing the diffusion process: $\langle r^2(t) \rangle = 6Dt$, we obtain an expression for the diffusion coefficient:

$$D(t) = 0{,}5025\, \pi^{4/3} n^{4/3} C_{DD}. \tag{64}$$

A comparison of (62) and (64) clearly reveals the weak sensitivity of the kinetics for the mean squared excitation delocalization in the crossover from the nondiffusion region to the diffusion region. These are, however, very strong changes for the diffusion coefficients.

It is possible to estimate the crossover time t_2 to diffusion kinetics if we set equal the first and second terms in square brackets in expression (60): $0.4393\, \pi n\, C_{DD}^{1/2} t_2^{1/2} = 1$; then

$$t_2 = 1/0.193\pi^2 n^2 C_{DD} \approx 29 W_F^{-1} = 14.5\, W_I^{-1}. \tag{65}$$

It should be noted that in spite of the substantial upper limit on this scheme with respect to concentration c and time t the authors of Ref. [114]

reproduced the general dynamics of the process with the excitation delocalization mechanism crossing over from a nondiffuse to a diffuse mechanism.

The authors of Ref. [115] attempted to overcome the limitations and nonrigorous nature of the calculation technique outlined above by using the diagrammatical technique to calculate a solution of the fundamental equation as a Green's function.

The authors carried out calculations for two- and three-site models, attempting to obtain expressions for the three primary parameters characterizing the migration of electron excitations: $I(E_L, t)$, $\langle r^2(t) \rangle$, and $D(t)$. The result included the detection of nondiffusion Förster excitation transport at the early migration stage (as discussed above) and diffusion transport for rather large times. The expressions obtained by the authors for the two limiting cases: the initial excitation transport stage ($t \to 0$) and the final stage ($t \to \infty$) made it possible to confirm and refine the results of Ref. [114].

Subsequent studies were devoted to a theoretical analysis of the same research on the dynamics of excitation transport in a disordered medium [116, 117]. These authors used theoretical models of continuous time random walks and an averaged Dyson equation for the theoretical analyses. The excitation migration process was generalized in these studies to the case of higher (than dipole-dipole) multipolarities of ion-ion interaction (dipole-quadrupole, quadrupole-quadrupole, etc.) and also compared the various theoretical models for describing electron excitation transport in disordered media.

Subsequent studies [118, 119] investigated disordered energy migration for both the four types of interaction (dipole-dipole ($S = 6$), dipole-quadrupole ($S = 8$), and quadrupole-quadrupole ($S = 10$) and exchange interaction) as well as the three spatial processes (one-, two-, and three-dimensional migration). It turned out that the solution for moderate and low concentrations became meaningless in the one-dimensional case for the shortest range ion-ion interaction mechanism (exchange interaction), i.e., the case of extremal excitation localization was recovered: $D = 0$. The diffusion stage remains relevant for two-dimensional migration, although it has a very small diffusion constant. Three-dimensional migration over substantial times is described by ordinary diffusion kinetics $\langle r^2(t) \rangle$.

Normal diffusion energy migration kinetics $\langle r^2(t) \rangle$ are observed for the multipole interaction mechanisms for all cases.

It is clear from table 1 that the diffusion coefficient D (taking reverse transfer into account) drops precipitously for all multipoles with diminishing dimensions of the migration process, although this is most strongly manifested for the "sharpest" interaction potential for $S = 10$. A drop in the diffusion coefficient of five to seven orders of magnitude results in a substantial enhancement of the quasilocalization of electron excitations in the one-dimensional case with quadrupole-quadrupole interaction.

Without examining the specific differences between the models and calculation methods outlined above we will provide results from the most thorough calculations of disordered migration which provide the fundamental picture of donor-donor energy transfer processes and are important from the viewpoint of their experimental analysis and comparison to theory.

1.*Small time and low concentration limit.* Nonstationary diffusion:

$$t_1 < t < t_2 = W_F^{-1} \tag{66}$$

For $S = 6$

$$\langle r^2(t) \rangle = 3{,}966\pi n\,[C_{DD}t]^{5/6} + 0{,}497\pi^2 n^2\,[C_{DD}t]^{4/3}, \tag{67}$$

$$D(t) = 0{,}548\pi n\,[C_{DD}t]^{1/6} + 0{,}0623\pi^2 n^2\,[C_{DD}t]^{1/3}, \tag{68}$$

$$I(E_L,\,t) = \exp[-(4/3)\pi n\Gamma(1-3/S)XC_{DD}t^{3/S}]. \tag{69}$$

2. *Large time and high concentration limit.* Stationary diffusion:

$$t < t_2 = W_F^{-1} \tag{70}$$

For random multipolarity of donor-donor interaction

$$\langle r^2(t) \rangle = 6\alpha(S)C_{DD}n^{(S-2)/3}t, \tag{71}$$

$$D = \alpha(S)C_{DD}n^{(S-2)/3}, \tag{72}$$

where $\alpha(S)$ is determined by the interaction multipolarity:

$$\alpha(S) = \frac{1}{2S}\,\frac{\Gamma(1-5/S)}{\Gamma(S/3)}\,\Gamma\!\left(\frac{5}{3}\right)\Gamma\!\left(1-\frac{3}{S}\right)^{(S-5)/3}\left(\frac{4}{3}\,\pi\right)^{(S-2)/3} \tag{73}$$

Table 1

Dependence of diffusion coefficient D (in relative units)
on the spatial type of the process and the interaction multipolarity

Type of process		S	
	6	8	10
Three-dimensional	0.4481	0.1309	0.251
Two-dimensional	0.1126	0.0150	0.0018
One-dimensional	0.0018	$1.3 \cdot 10^{-5}$	$6.2 \cdot 10^{-8}$

It is important that the interrelationship obtained in this case

$$\langle R^2(t) \rangle = 6Dt \tag{74}$$

precisely corresponds to the solution for excitation diffusion along an ordered lattice, where introduction of the diffusion coefficient is strictly justified (Formula (69) was obtained in Refs. [111, 112] while the remaining were obtained in Ref. 117]).

The functional dependence of the diffusion coefficient in the disordered medium on the interacting particle concentration is important; this relation is determined by Expression (72) for different multipolarities of donor-donor interaction.

As we discussed above such a form of the relation ($D \sim n^{4/3}$) for dipole-dipole interaction was first obtained in Ref. [23] which introduced the concept of the quasidiffusion coefficient for describing electron excitation transport in a disordered vitreous medium.

A comparison of the result predicted in Ref. [119] to solution (72) for dipole-dipole interaction obtained in Refs. [113, 115-117] shows that the nature of the functional relation $D \sim n^{4/3}$ is identical in all cases and the only difference appears in the numerical value of the coefficient α in the expression for D.

Unlike the solutions for $\langle R^2(t) \rangle$ and $D(t)$ where the majority of authors have obtained identical functional relations that differ only insignificantly in their coefficients substantially different results have been obtained from using different calculation models [117] for the third, and to us, the most important quantity $I(E_L, t)$ which can be measured experimentally. If analytic expression (69) has been reproduced by a number of authors for small t and c [115, 118] and has been tested by computer modeling [111, 112] there remains no uniform opinion regarding the decay law of $I(E_L, t)$ for large t and c (the diffusion stage). In the majority of studies [112, 115-118] the calculation results for $I(E_L, t)$ are given as Laplace transforms without inversion on the time axis, which requires cumbersome numerical techniques. This significantly complicates the analysis and comparison of functional relations calculated by different methods.

The similarity of Solutions (71) and (74) for a disordered lattice to solutions for excitation diffusion throughout an ordered lattice suggests that the population decay kinetics of selectively-excited centers are similar and make it possible to apply the solution for the ordered case (see Expression (41)) to the disordered case:

$$I(t) = (4\pi n^{2/3}Dt)^{-3/2} \tag{75}$$

Such a solution for the diffusion decay stage where the population profile is stationary can be found from Expressions (71) and (74). Normalizing the complete number of excitations in a sphere of radius $L(t) = \sqrt{\langle r^2(t) \rangle}$ to unity and substituting the true distribution $\psi(r, t)$ with a rectangular distribution:

$$\psi(r, t) = \begin{cases} \psi(0, t) & \text{for } r < L(t) \\ 0 & \text{for } r > L(t), \end{cases}$$

we obtain for the excitation density per unit volume

$$\psi(0, t) = L^{-3}(t) = (6Dt)^{-3/2}. \tag{76}$$

Hence the excitation density per ion takes the form

$$I(t) = \psi(0, t)/n = (6n^{2/3}Dt)^{-3/2}, \tag{77}$$

accurate to a coefficient analogous to Expression (75).

Because the spatial and spectral distributions of the excited particles are statistically independent the requirement for a stationary spatial population profile does not infer its spectral stationarity, which makes it possible to follow $I(t)$ using spectral selection of these centers.

The expression derived for the diffusion in a three-dimensional medium [77] has two- and one-dimensional analogs:

$$\psi(0, t) \approx (Dt)^{-1} - \text{migration along a planar grid} \tag{78}$$

$$\psi(0, t) \approx (Dt)^{-1/2} - \text{migration along a linear chain} \tag{79}$$

A comparison of Expressions (76), (78) and (79) clearly reveal that the more complex two- and three-dimensional solutions can be obtained by simply raising the one-dimensional solution to the corresponding power ($d = 2$ and $d = 3$).

Ref. [120] is devoted to a strict analysis of the kinetics of migration along a one-dimensional chain with randomly-distributed breaks. It turned out that the exact solution contained, in addition to the initial exponential behavior and diffusion law $t^{-1/2}$, a second exponential stage in the intermediate range over long decay times:

$$I \approx \exp[-(\lambda t)^{1/3}]. \tag{80}$$

This stage is directly related to the excitation localization effect at isolated sections along the linear chain.

A more detailed analysis of this effect [120] demonstrated a strong dependence of the type of decay kinetics $I(t)$ as $t \to 0$ on the distribution function of the ion-ion interaction probabilities $\rho(W)$ near $W \to 0$.

Without dwelling on the details we provide expressions for $I(t)$ for the most important decay cases in the late stage $t \to \infty$ for different probability distributions $\rho(W)$.

1. Zero localization (no breaks in the chain):

$$\rho(W) \to 0 \text{ for } W \to 0, \tag{81}$$

$$I(t) \approx (4\pi W_{\text{eff}})^{-1/2} t^{-1/2}, \tag{82}$$

where $W_{\text{eff}}^{-1} = \int (1/W)\rho(W)dW$ (see (79)).

II. Weak quasilocalization (few breaks):

$$\rho(W) \to \text{const} \neq 0 \text{ for } W \to 0, \tag{83}$$

$$I(t) \sim (\ln t)^{1/2} t^{1/2}. \tag{84}$$

III. Strong quasilocalization:

$$\rho(W) = W^{-q} \; (q < 1) \text{ with the extremum as } W \to 0 \tag{85}$$

$$I(t) \sim t^{-(1-q)/(2-q)}. \tag{86}$$

IV. Extremal localization:

$$\rho(W) = p\delta(W) + (1 - p)\delta(W - W_0), \tag{87}$$

where p varies from 0 to 1. The existence of the distribution function $\rho(W)$ as $W \to 0$ creates localization and here

$$I(t) = (1/\sqrt{4\pi W_0 t}) \, [1 + (\lambda t)^{1/3}] \, \exp \, [-(3/2) \, (\lambda t)^{1/3}], \tag{88}$$

where $\lambda = \frac{1}{2}\pi^2 q^2 W_0$, $q = -\ln(1 - p)$; W_0 is the rate of excitation transfer over the minimum possible distance.

Unlike Expressions (82), (84), (86) corresponding to the late stage of the decay kinetics Expression (88) is exact for all kinetics $I(t)$ and for $r = 0$ when all couplings are made (no breaks), and which includes the diffusion stage (Expression (82)) as the final stage.

We emphasize that the authors of Ref. [120] determine the expression for the effective rate in the diffusion stage by averaging the inverse transfer probability or the hop time rather than by averaging the probability. This result confirms those obtained in Ref. [94] from a calculation of excitation migration along a linear chain of spectrally-disordered donors to the acceptor.[4]

With a small yet finite number of breaks in the chain p is small, yet not equal to zero, and the diffusion kinetics (82) are observed only in the intermediate times when

$$1 \ll 2W_0 t \ll 4/\pi^2 q^2.$$

Here in the late stage there will always be an exponential decay characterizing the extremal excitation localization:

$$I(t) \sim \exp[-3/2(\lambda t)^{1/3}]. \tag{89}$$

Speculating with regard to the two- and three-dimensional nature of diffusion in the plane and volumetrical cases, respectively, we can roughly apply Expression (88) to these by raising all time-containing terms to the second and third powers, respectively, as was done previously for diffusion

---

[4] As demonstrated in Ref. [13] the formal and unsubstantiated incorporation of inhomogeneously broadening in migration by averaging the transfer probabilities yields qualitatively incorrect results such as no maximum for the effective migration rate W where homogeneous and inhomogeneous broadening is equal ($\delta = \Delta$).

throughout an ordered lattice (see (77), (79)). In this case we obtain for three-dimensional diffusion which is more interesting from the experimental viewpoint:

$$I(t) \approx (4\pi W_0 t)^{-3/2} (1 + \lambda t) \exp \left[(-3/2)^3 \lambda t \right] . \tag{90}$$

As we see this relation is similar to Expressions (75)-(77) and, like (88) determines the diffusion decay kinetics prior to excitation localization

$$t_{\mathrm{L}} = \frac{4/\pi^2 q^2}{2 W_0} \tag{91}$$

as well as the exponential kinetics following this time.

These arguments are clearly approximations and do not account for differences in the nature of diffusion or localization of excitations in media with different dimensionalities (one-, two-, and three-dimensional cases) and various degrees of short-range action for energy transport ($S = 6, 8, 10$, see Table 1).

Another approach to examining spectral diffusion with a random impurity distribution in a three-dimensional lattice to find an expression for $I(t)$ was provided in Ref. [121]. The authors sought a solution for $I(t)$ in the remote and intermediate cases of excitation decay using an analogy between the diffusion problem and the problem of current pulse propagation in a three-dimensional network with random impedances. Citing the lack of results from an analytic analysis of $I(t)$ in the literature for a three-dimensional disordered system the authors of [121] carried out the necessary calculation with a type (87) transfer probability distribution which was used previously in the one-dimensional model. After analyzing a Bethe lattice having k bonds for each particle and a locally-similar real crystal with the coordination number $z = k + 1$ (k is the degree of branching) they determined $I(t)$ for two regions: above and below the percolation threshold

$$c_{\mathrm{T}} = (z - 1)^{-1} = 1/k. \tag{92}$$

These solutions can be given as:

$$I(t) = B_1 + \sum_{n=1}^{k+1} B_2^{(n)} \left[1 + B_3^{(n)} (W_0 t/a)^{n-1} \right] \exp\left(-W_0 \left| \Omega_0 \right| t/a \right) \tag{93}$$

for $c \to c_{\mathrm{T}} + 0$, $W_0 t/a \gg 1$;

$$I(t) = B_4 + B_5 \exp\left(-W_0 \left| \Omega_1 \right| t/a \right) + \sum_{n=1}^{k+1} B_6^{(n)} (W_0 t/a)^{n-1} \exp\left(-W_0 \left| \Omega_2 \right| t/a \right) \tag{94}$$

for $c < c_{\mathrm{T}}$, $W_0 t/a \gg 1$. Here Ω_0, Ω_1, Ω_2 are dimensionless parameters dependent on the concentration c.

We will not provide the expressions for the coefficients B_1, ..., B_6

due to their cumbersome size, but rather will discuss only the exponents in which

$$a = c_T Q^k - 1/\epsilon^2 = c_T Q^k - 1/(1 - c_T k Q^{k-1})^2, \tag{95}$$

where in turn

$$Q = 1 - \frac{2(c - c_T)}{1 - c_T} + O(c - c_T)^2 + \dots \tag{95}$$

In the late decay stage time dependences $I(t)$ for both cases (93) and (94) take the form

$$I(t) \sim (W_0 t/a)^k \exp(-W_0|\Omega_i|t/a), \tag{97}$$

where

$$|\Omega_i| = |\Omega_0| = 0.603 \text{ as } c \to c_T + 0$$
$$|\Omega_i| = |\Omega_2| = 0.7301 \text{ for } c < c_T.$$

It is clear from (93) that the decay rate on the kinetic curve $I(t)$ above the percolation threshold was slower than below this threshold.

For the intermediate time range the solution takes the form

$$I(t) \sim 1/\sqrt{W_0 t}, \; c \to c_T, \; \epsilon \ll W_0 t/a \ll 1, \tag{98}$$

recalling the diffusion solution for a linear chain.

This led the authors to propose that as $c \to c_T$ the Bethe lattice became a composition of linear chains in which the branch points and intersections of the chains played no substantial role. However, we note that such kinetics ($\sim 1/\sqrt{t}$) are substantially different from the diffusion kinetics for an ordered Bethe lattice ($c = 1, k \neq 1$) for which the authors of Ref. [121] obtained a somewhat dubious exponential decay:

$$I(t) = \frac{k(k+1)}{k^2+1} \exp\left[-\frac{(k-1)(k^2-1)}{k^2+1} W_0 t\right]. \tag{99}$$

Noting that their method is universal for any transport rate distributions and having obtained a good agreement between their solution for the nonlinear case and the known result (84) the authors cite the mathematical complexities of deriving an exact result for a weakly-disordered Bethe lattice ($c \to 1$). This makes it impossible to evaluate the reliability of Expression (99) for an ordered Bethe lattice which is substantially different from the existing solution for diffusion throughout an ordered lattice (39).

An analysis of cases I–IV above suggests that some progress can be made in understanding the degree of localization (or quasilocalization) of excitations in actual disordered media by precise measurement of the specific form of the late stage of the excitation decay kinetics $I(t)$ and by comparison to calculation results using various theoretical models.

We know of two studies that employed the continuous-time random walk (CTRW) technique to obtain an analytic form of the solution for the diffusion stage of disordered migration [119, 122]. Ref. [119] obtained a solution for three-dimensional migration with multipole interaction

$$I(t) \sim (W_{\text{diff}}t)^{-3/2}, \tag{100}$$

that is similar to the solution for ordered diffusion and also provides an estimate for the average disordered diffusion rate [89, 90], which was equal to the average migration rate in the nonstationary stage:

$$W_{\text{diff}} = W_I = [(4/3)\pi n \Gamma(1 - 3/S)X]^{S/3}C_{\text{DD}}. \tag{101}$$

Unfortunately, the cited study [119] contains no discussion of the accuracy or the limits on the applicability of this asymptotic solution as well as no comparison to results of other models as $t \to \infty$. It should also be pointed out that the very applicability and possible accuracy of the CTRW technique for application to this migration problem have been discussed in the literature, although no test of the technique for calculating the nonstationary kinetics (where a rather exact solution is known) (see Expression (71)) has been carried out in Ref. [119].

Refs. [122-124] propose a semi-phenomenological approach for describing migration throughout a disordered medium. Introducing a singularity at the excitation start point in the CTRW theory the authors construct approximate solutions of the resulting equations by joining the asymptotic solutions for small ($t \to 0$, $I \sim 1 - \sqrt{W_L t}$) and large ($t \to \infty$, $I \sim t^{-3/2}$) times:

$$I(t) = \exp(-\sqrt{W_F t}) + [1 - \exp(-\sqrt{W_F t}) \, \{1 + 1{,}93 \, [0{,}79 \, W_F (t + \tilde{\tau})]^{-1/2}\} \times$$
$$\times \, [0{,}79 \, W_F (t + \tilde{\tau})]^{-3/2}, \tag{102}$$

where $\tilde{\tau} = 4.58 \, W_F^{-1}$. The accuracy of this as well as the preceding approach lies beyond the scope of the present analysis.

One feature of this solution that was highlighted by the authors of Refs. [122-124] is the characteristic overvibration between the initial nonstationary and long-term diffusion stages of the process. This vanishes in the transition to interaction with potentials that are "sharper" than the dipole-dipole interaction.

It is clear form Expression (102) that this solution is based on Förster irreversible energy transfer kinetics ($I \sim \exp(-\sqrt{W_F t})$). At early stages when $W_F t \ll 1$, $I(t) > 0.37$ when the series expansion of the exponent to the first term holds, the reversibility of energy transfer is accounted for by introducing the coefficient $\tilde{\tau}$ in the second term. With increasing $t \gg W_F^{-1}$ and diminishing $I(t) < 0.37$ such a process of accounting for reversibility becomes unsatisfactory and the decay kinetics become Förster irreversible transfer kinetics (see Expression (30)) which increasingly deviate in intensity and rate from the reversible transfer kinetics in the Huber pair back transfer model [111, 112] (see Expression (51)). Even for $I = 0.1$ the differences in intensity in the Huber [111] and Dzheparov [122-124] models ~35% and then grow with increasing t and diminishing $I(t)$. It is possible to assume that improper incorporation of transfer reversibility (back to the initial center) in the Dzheparov model is in fact what causes the crossover in the intermediate stage from slow Huber kinetics to the faster Förster kinetics and then on to the slower diffusion stage, demonstrating strong overvibration. The exaggerated values of the diffusion coefficient D and the numerical coefficient $\alpha = 3.46$ in the Dzheparov model compared to all other theoretical models also indicate poor incorporation of transfer reversibility in the stationary kinetic diffusion stage ($t \rightarrow \infty$).

The issue of finding the exact complex multistage decay kinetics of the initial centers resulting from disordered migration has therefore remained until recently. The theoretical and experimental principles and the initial solutions of this problem are discussed in detail in Ref. [125].

4. Migration-Controllable and Rapid Quenching Energy Transfer Among Donors and Acceptors (Theoretical Models)

In Section 2 we considered the case where high acceptor concentrations are required for deactivation of electron excitations (donor quenching by acceptors), which results in rather small distances R_{DA} at which direct excitation transfer from donor to acceptor takes place. There is, however, another, no less important, case where excitation (quenching) transfer occurs at exceedingly small concentrations n_A that are clearly insufficient for direct donor-acceptor interaction.

This case was first identified in Ref. [126] from an investigation of luminescence quenching in anthracene crystals by introducing a small concentration of naphthacene. It turned out that donor (anthracene) excitation deactivation resulted from the excitation energy migration among the excited states of the donor molecules to the quencher (naphthacene) molecules. This quenching case most often occurs as soon as $n_D \gg n_A$, since this requires, in addition to static donor-acceptor energy transfer, effective donor-donor interaction resulting in migration to the quencher.

The mathematical problem of excitation migration is similar to the familiar problem of actual particle diffusion that occurs in, for example, liquid media with the difference that the diffusion coefficient here is not determined by particle dimensions, viscosity, or temperature, but rather by the donor-donor interaction probability.

The diffusion equation for describing energy migration is often written as[5]

$$\partial P(r, t)/\partial t = D\Delta P(r, t), \tag{103}$$

where $P(r, t)$ is the donor excitation density; D is the diffusion coefficient:

$$D = \frac{1}{6}\sum_j W_{ij}r_{ij}^2 = \frac{1}{6}C_{DD}\sum_j r_{ij}^{-4}, \tag{104}$$

C_{DD} is the donor-donor interaction ($S_{DD} = 6$).

The solution of this equation subject to the boundary condition $P(r, t)|_{r=a} = 0$ results in donor excitation decay kinetics of the following type:

$$I(t) = P(t)/P(0) = \exp(-\overline{W}t - \overline{W}a\sqrt{t/\pi D}). \tag{105}$$

Here the first term in the exponent $(-\overline{W}t)$ is responsible for the exponential kinetics of donor luminescence and corresponds to the stationary solution of the diffusion equation. The migration quenching probability $\overline{W}$ is linearly dependent on acceptor concentration n_A, diffusion coefficient D, and the characteristic excitation scattering length (the radius of the quenching sphere) a:

$$\overline{W} = 4\pi n_A Da. \tag{106}$$

The second term in the exponent corresponds to the nonstationary solution of the diffusion equation and results in a nonexponential initial section of the donor excitation decay curve. This nonexponential nature ordinarily is manifested only under conditions of small diffusion coefficients [49, 50, 128].

This situation corresponds to the case of a "black quenching sphere" when interaction with the acceptor is accounted for only in the boundary condition.

The diffusion approach is strictly substantiated only in the case of a regular donor configuration, i.e., for $n_D \approx 100\%$. Strong fluctuations arise in the donor-donor interaction probability W_{if} for dilute impurity crystals which complicates the introduction of a single diffusion coefficient. However, analysis of theoretical studies and our experimental results described previously (see Section 5) make it possible to write the following expression for the diffusion coefficient through a disordered medium taking the irregularity of the donor configuration into account

$$D = \alpha n_D^{4/3}C_{DD}, \tag{107}$$

where α is the numerical coefficient which varies in the different theoretical

---

models. We will discuss this coefficient in some detail in our analysis of experimental data on reverse migration.

Substituting expression (107) for D into (106) we obtain

$$\overline{W} = 4\pi\alpha n_{\mathrm{D}}^{4/3} n_{\mathrm{A}} C_{\mathrm{DD}} a. \tag{108}$$

It is clear that if the radius of the quenching sphere is a fixed quantity the linear relation between the migration quenching rate $\overline{W}$ and C_{DD} will appear only when there is an "ultralinear" dependence on donor concentration: $\overline{W} \sim n_{\mathrm{D}}^{4/3}$.

The boundary condition may be important in diffusion problems. In actual situations the radius of the quenching sphere will not always be fixed. Thus, as the diffusion coefficient D changes the dimensions of the area in which direct quenching competes with migration may change. Formally this can be accounted for through the boundary condition if the radius of the quenching sphere is defined as the distance over which the direct quenching rate and the migration rate through the quenching sphere are equal, i.e., $W_{\mathrm{DA}}(a) = \tau_{\mathrm{M}}^{-1}(a)$, where $W_{\mathrm{DA}}(a) = C_{\mathrm{DA}}/a^6$; $\tau_{\mathrm{M}} = a^2/D$ is the diffuse propagation time of a sphere of radius a, i.e.,

$$a = (C_{\mathrm{DA}}/D)^{1/4}. \tag{109}$$

Essentially this is a rectangular approximation of the quenching center. It is clear that a radius introduced in this manner will depend on C_{DA} and the diffusion coefficient D. Substituting Expression (109) into (106) we obtain for the migration probability

$$\overline{W} = 4\pi n_{\mathrm{A}} D^{3/4} C_{\mathrm{DA}}^{1/4}. \tag{110}$$

A solution of the more general problem when the terms accounting for interaction with the acceptors is directly introduced into the stationary diffusion equation (and not through the boundary conditions) is provided in monograph [129]. For the dipole–dipole donor–acceptor quenching mechanism this equation takes the form

$$D\Delta n_{\mathrm{D}}^* - \frac{1}{\tau_{\mathrm{D}}^0}\left(\frac{R_0}{r}\right)^6 n_{\mathrm{D}} + 1 = 0. \tag{111}$$

The expression for the excitation lifetime taking the spontaneous decay time τ_{D}^0 into account was obtained by a method analogous to the Wigner–Seitz cell technique:

$$\tau_{\mathrm{D}} = \tau_{\mathrm{D}}^0 f\left(\frac{R_0}{\sqrt{D\tau_{\mathrm{D}}^0}}, \frac{\sqrt[3]{3n_{\mathrm{A}}/4\pi}}{R_0}\right), \tag{112}$$

where $f\left(\dfrac{R_0}{\sqrt{D\tau_D^0}}, \dfrac{\sqrt[3]{3n_A/4\pi}}{R_0}\right)-$ is a dimensionless function given in Ref.

[129] for a number of values of the parameter $R_0/\sqrt{D\tau_D^0}$ and $\sqrt[3]{3n_A/4\pi}/R_0$.

This table was republished in Ref. [130] with a more exact form of the

function $P(r)$ for $R < R_0$ and for

A solution of Equation (111) gives us an expression for the donor deactivation probability $\bar{W}_D = 1/\tau_D - 1/\tau_D^0$ describing the exponential asymptotics of donor decay at the late time stages when quenching is determined by stationary excitation diffusion from rather remote regions from the acceptor.

However, as we have observed previously, the dipole–dipole donor–acceptor interaction also is directly responsible for the absolute decay of excitations near the acceptor (excluding migration). Competition between these two processes may result in significantly more complex decay kinetics along the entire time scale. The joint effect of direct transfer and particle diffusion on the kinetics of donor luminescence decay was first examined in Ref. [131]. The authors solved the complete temporal diffusion equation by means of a series expansion in the parameter D:

$$\partial P(r,\,t)/\partial t = -P(r,\,t)/\tau_D^0 + D\Delta P(r,\,t) + \sum_i W_{DA}(|r - r_i(t)|)P(r,\,t). \tag{113}$$

The kinetics were complex so that in the early decay stage they are similar to the static Förster law $(\sim e^{-\gamma_F\sqrt{t}})$ while in the final stage they become exponential as $e^{-\bar{W}t}$ with the exponent

$$\bar{W} = 0{,}68 \cdot 4\pi n_A D^{3/4} C_{DA}^{1/4}. \tag{114}$$

Substituting Expression (107) for D into this expression we obtain

$$\bar{W} = 0{,}68 \cdot 4\pi\alpha^{3/4} n_D n_A C_{DD}^{3/4} C_{DA}^{1/4}. \tag{115}$$

As we see, this refined solution accurate to the constant 0.68 coincides with the estimate of the rate of dipole–dipole migration to dipole quencher (110) and demonstrates the linear dependences on donor and acceptor concentrations with suppressed dependences on C_{DD} and C_{DA}.

The dependence of migration rate on donor concentration and the microparameters becomes increasingly complex for the case of donor–donor and donor–acceptor interactions with multipolarities S_{DD}, S_{DA} of 6, 8, and 10 [90]:

112   *Alimov et al.*

$$\bar{W} \sim C_{DD} n_A n_D^{S_{DD}/3}, \tag{126}$$

and specifically, for $S = 6, 8, 10$

$$\bar{W} \sim n_D^2; \ n_D^{8/3}; \ n_D^{10/3}. \tag{127}$$

In actual media (crystals and glasses) there is always a short-range order of the first and second coordination shells, unlike the continuous media examined above; at high donor concentrations ($n_D \gg n_A$) this will manifest itself analogous to the short-range, contact quenching mechanism since the quenching probability will not vary gradually as the acceptor is approached, which is the case in a continuous medium, but rather in a hop (when excitation is transferred from the donor in the second coordination shell around the acceptor to the donor in the first shell). Such interaction and structural features of solids may also result in stronger than linear dependences of the migration rates on donor concentration.

Refs. [25, 132, 133] have also been devoted to an investigation of the excitation decay kinetics with random C_{DD}, C_{DA}, n_D and n_A. These studies analyze complete kinetic equations that account for the migration and quenching of donor excitations. In spite of the various approaches to the problem and the different final results we will discuss the common aspects.

The kinetics are written as an exponent consisting of the sum of two terms (the expressions are not provided here in view of their cumbersome size). One term describes the static Förster donor quenching process while the other is due to the "mixing" action of excitation migration throughout the donor system. The expressions obtained in these studies for the exponential decay factor in the late stages in the hop migration limit are in good agreement with one another and with Expression (122) obtained in Ref. [23].

An analysis of strong interaction for the hop excitation migration mechanism was carried out in Refs. [22, 134]. The maximum quenching rate with ultrafast migration is limited by energy leakage in the lattice. In this case it is no longer dependent on C_{DD} and takes the following form for irreversible dipole-dipole donor-acceptor interaction [22]

$$W_{max} = (2/3)\pi^2 n_A C_{DA}/R_{min}^3. \tag{128}$$

Since it is exceedingly difficult to obtain an exact solution of the multiparticle problem with random concentrations n_D and n_A and account for donor-donor, acceptor-acceptor, and donor-acceptor interactions Ref. [135] analyzed the applicability conditions of the collective kinetic equations for describing interaction between two impurity subsystems. It turned out that they were not only applicable but also convenient for the strong interaction case.

Another approach considered by the authors to be a more general routine for solving the problem of energy transfer from donors to acceptors accounting for donor-donor and donor-acceptor interactions was developed in Refs. [28, 111]. This approach is based on an analysis of a group of coupled rate equations for the donor excitations. The donor luminescence decay kinetics $I(t)$ are sought by the authors as Laplace transforms

$$I(s) = \int_0^\infty dt\, e^{-st} I(t). \tag{129}$$

Using an approximation of the average T-matrix (for $c \ll 1$) the author obtains the result:

$$I(s) = \left[s + c_A \sum_{l,l} T_{l,l'}(s) \right]^{-1}, \tag{130}$$

where the T-matrix is a solution of the equation

$$T_{l,l'} = X_{0,l}\delta_{l,l'} - \sum_{l''} X_{0,l} \langle g_{l,l''}(s) \rangle T_{l'',l'}(s), \tag{131}$$

in which $X_{0,l}$ is the donor-acceptor transfer rate, while $\langle g_{l,l'}(s) \rangle$ is the configurational average of the Green's function for donor-donor energy transfer without acceptors.

In the fast energy transfer case ($W_{ij} \to \infty$) using the Born approximation ($T_{l,l'} \to X_{0,l}\delta_{l,l'}$) the authors of Refs. [28, 111] obtained a result for the kinetic limit $I(s) = [s + c_A \sum_l X_{0,l}]^{-1}$ which coincides with existing expressions from Refs. [22, 90, 134].

In the weak donor-donor transfer case ($W_{ij} \to 0$)

$$g_{l,l'}(s) = \delta_{l,l'} s^{-1}, \quad T_{l,l'}(s) = \frac{\delta_{l,l'} X_{0,l}}{1 + X_{0,l}/s},$$

$$I(s) \approx s^{-1} - c_A s^{-1} \sum_l X_{0,l}(s + X_{0,l})^{-1},$$

which corresponds to the static quenching kinetics

$$I(t) \approx 1 + c_A \sum_l (l^{-X_{0,l}t} - 1).$$

In the general case with a random W_{ij} Expression (130) also permits reproduction of the results of both migration models: the diffusion and hop models.

Therefore, in a continual approximation going from the sum to the integral in (130) and (131) it is possible to arrive at the diffusion model by writing for $g_{l,l'}(s)$

$$g(r, r', s) = (2\pi)^{-3} \int dk \exp[2k(R - R')] (s + Dk^2)^{-1}. \tag{132}$$

Here D is the diffusion coefficient which takes the following form for a donor lattice with translational symmetry

$$D = \frac{1}{6}\sum_j W_{ij}(R_i - R_j)^2 . \tag{133}$$

As noted above it is quite difficult to introduce and calculate diffusion coefficients for a disordered system ($c \ll 1$). Using the concepts of the dimensions and average distance between donors Huber in Ref. [111] provides an approximate expression for D:

$$D = \alpha(S)C_{DD}n_D^{(S-2)/3} = [2(S-5)]^{-1}\left(\frac{4\pi}{3}\right)^{(S-2)/3} C_{DD}^{(S-2)/3} n_D, \tag{134}$$

which reproduces the result of Ref. [23] for $S = 6$ and differs from the more exact calculations (see (74), (75)) by the value of the coefficient $\alpha(S)$: $\alpha(6) = 3.38$; $\alpha(8) = 2.93$; $\alpha(10) = 4.55$.

The results for the hop model were obtained in Ref. [111] from Expression (130) when the nondiagonal elements in the T-matrix are neglected:

$$I(s) = \{ s + c_A \sum_l X_{0,l}/[1 + X_{0,l}g_0(s)]\}^{-1}, \tag{135}$$

where $g_0(s) = g_{ll'}(s)$.

Relation (135) is identical to the result obtained for Burshteyn hop model (122) if we write $g_0(s) = (s + 1/\tau_o)^{-1}$, $g_0(0) = \tau_0$.

However, a more exact $g_0(s)$ is a Laplace transform of the initial excitation decay kinetics in selective laser excitation experiments (see Section 5) for $c_A = 0$;

$$g_0(s) = \int_0^\infty dt e^{-st}I(E_L, t) . \tag{136}$$

For the case of a diluted donor system ($c \ll 1$) and using Expression (71) for $I(E_L, t)$, $I(E_L, t) = \exp(-\gamma_I t^{3/S})$, we obtain

$$g_0(0) = \tau_0 = \Gamma(S/3 + 1)\gamma_I^{-S/3},$$

where the reversible donor energy transfer macroparameter

$$\gamma_I = (4/3)\pi n_D C_{DD}^{3/S}2^{3/S-1}\Gamma(1 - 3/S).$$

Applying the inverse Laplace transform to Expression (135) yielded exponential decay kinetics over long time periods ($t \to \infty$, $s \to 0$):

$$I(t) = \exp\left[-c_A t \sum_l \frac{X_{l,l}}{1 + X_{0,l}g_0(0)}\right]. \tag{137}$$

This expression can be given as

$$I(t) = \exp\left[-c_A t \sum_l (X_{0,l}^{-1} + \tau_0)^{-1}\right], \tag{138}$$

where the generalized donor excitation decay probabilities $(X_{0,l}^{-1} + \tau_0)^{-1}$ are summed and in the two limiting cases these are determined solely by the excitation delivery rate by the donors $(X_{0,l}^{-1} \ll \tau_0)$ or the rate of transfer to the acceptor $(X_{0,l}^{-1} \gg \tau_0)$ (analogous to Expression (128)).

The conditions for a transition to integration (or neglecting nondiagonal terms in the T-matrix) yield the boundary conditions on the applicability of the diffusion and hop models:

$$n_D^{1-S_{DD}/S_{DA}}(C_{DA}/C_{DD})^{3/S_{DA}} \gg 1 - \text{diffusion model}$$

$$n_D^{1-S_{DD}/S_{DA}}(C_{DA}/C_{DD})^{3/S_{DA}} \lesssim 1 - \text{hop model}$$

Recognizing the limited application of Expressions (130)-(138) derived above for $I(s)$ the case $c_A \ll c_D$, Huber used a coherent potential approximation in Ref. [136] and applied the results to the case of a random c_A/c_D ratio. This can be successfully employed only for dipole-dipole donor-acceptor and donor-donor interactions $(S_{DA} = S_{DD} = 6)$ for $c_A \leqslant 1$ and $c_D \ll 1$ neglecting the nondiagonal terms in the T-matrix, i.e., for $C_{Da} \leqslant C_{DD}$.

The general solution in this case takes the form

$$I(s) = [s + X_{c.p.}(s)]^{-1}, \tag{139}$$

where $X_{c.p.}(s)$ is the solution of the nonlinear self-consistent equation. In the limit $t \to \infty$ $(s \to 0)$

$$I(t) \sim \exp[-X_{c.p.}(0)t]. \tag{140}$$

The analytic expression for $X_{c.p.}(0)$ can be written only in the low concentration limit $(c_A \ll c_D)$:

$$\overline{W} = X_{к.п}(0) = (4/9)\pi^{7/2} n_A n_D \sqrt{C_{DD}C_{DA}}. \tag{141}$$

It is this case $(c_A \ll c_D)$ that has been repeatedly analyzed in the past (see Expression (122)) yielding identical results accurate to the numerical coefficient.

The behavior of $X_{c.p.}(0)$ for large c_A is shown in Figure 2 from which it is clear that the linear approximation of $X_{c.p.}(0)$ begins to function poorly as early as $n_A\sqrt{C_{DA}}/n_D\sqrt{C_{DD}} \geqslant 0.5$. Such a strong deviation of the concentration dependence of the deactivation rate from a linear dependence had not been predicted by other theories and has not yet been observed experimentally. This requires additional theoretical confirmation as well as formulation of new experiments for special analysis of this issue.

Expression (141) was obtained by integration of kinetic equations with respect to R within the limits from 0 to ∞ which is equivalent to an infinitely large quenching rate $(W_{DA} \rightarrow \infty)$ at a zero intervening distance between the systems at the same time that the actual system and calculation of the kinetic limit (fast migration) require establishing the lower integration limit $(R_{min} \neq 0)$, which places an upper limit on the rate W_{DA}: $W_{DA}^{max} = C_{DA}/R_{min}^6$. Here relation (141) becomes

$$X_{\text{к.п}}(0) = (4/9)\pi^{7/2} n_A n_D \sqrt{C_{DA} C_{DD}} \, [1 - (2/\pi)\text{arctg}(R_{min}^6/C_{DA}\tau_0)^{1/2}] . \quad (142)$$

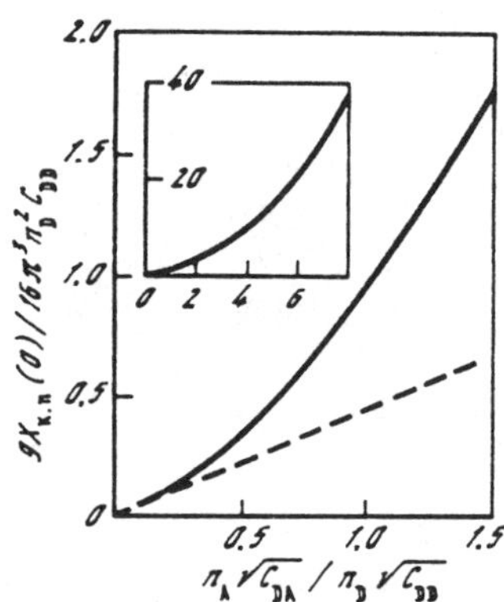

Fig. 2. Behavior of the normalized migration rate $X_{\text{c.p.}}(0)$ at large acceptor concentrations c_A.

The second term in square brackets here has no substantial effect on $X_{\text{c.p.}}(0)$ which coincides with expression (141) for $(R_{min}/C_{DA}\tau_0)^{1/2} \ll 1$. The latter condition is equivalent to a low average rate of excitation transport to the acceptor compared to its maximum relaxation rate at the acceptor: $\tau_0^{-1} \ll C_{DA}/R_{min}^6$ (migration-controllable quenching).

In the opposite case of fast excitation migration to the acceptor which saturates donor-acceptor relaxation $(\tau_0^{-1} \ll C_{DA}/R_{min}^6)$ the analyzed expression is reduced to (128) where the dependence on the migration rate among the donors and their interaction efficiency C_{DD} (the kinetic limit) vanishes.

5. Dynamics of Electron Excitation Migration in Disordered Media (Experiment)

The most important form of collective interactions directly responsible for a wide range of electron excitation transport and relaxation processes in solids are donor-donor migration interactions of active ions whereby electron excitation will only alter their position in the crystal resulting in zero or near-zero energy losses.

The studies of electron excitation migration kinetics in disordered solids, the processes of establishing the micromechanisms and determining the efficiency of elementary donor-donor interactions which are fundamental issues in solid state physics are becoming increasingly important from a practical viewpoint today. This is due to the broad utilization of quantum amplifiers, laser oscillators and anti-Stokes converters in quantum electronics; such devices employ active elements with high concentrations ($\sim 10^{21}$ cm^{-3}) of the working impurity (see Studies [4, 6-8] and the references contained therein). Energy migration plays a very important role in such systems and is responsible not only for substantial enhancement or degradation in the efficiency of the device, but also to some degree determines its very functional capability.

One traditional method of investigating excitation migration is an indirect method of analyzing concentration luminescence quenching [1, 2, 5, 9-12, 14, 15, 22] (see also Section 4) where migration among the donors determines excitation transfer to the quencher (acceptor). Emphasizing the scientific and practical importance of analyzing this phenomenon we recognize its limitations as a method of investigating migration due, as we have seen, to the broad variety and complexity of its theoretical description [20, 22, 28, 89].

In this section we analyze electron excitation migration processes in pure form, i.e., when the medium does not contain any extraneous quenchers (energy acceptors). There are several methods for direct observation of energy migration in existence today. One involves analyzing the depolarizing action of migration on the luminescence excited by polarized light [137, 138]. Other techniques use the capability to implement spectral monitoring of migration in the case of inhomogeneous line broadening. These include "hole burning" in the luminescence line under induced relaxation and an investigation of such relaxation [97-102, 139-142], observing the "red shift" of the luminescence lines (in the stationary [30, 143-145] and kinetic [6, 146] variants) and excluding monochromatic excitation due to the specific selectivity of low-temperature energy transfer.

This group of techniques includes applying selective laser excitation methods to investigating energy migration as proposed in Ref. [147] and developed in a number of studies [8, 18, 31, 84, 148]. The latter method (selective laser spectroscopy) in its kinetic (nonstationary) variant appears to be the most convenient and valuable from the viewpoint of information recovered, since it does not require satisfying oscillation or amplification conditions in the test material and has no temperature limitations, as do those examined previously. By selecting laser excitation sources this method can be used to investigate migration among rare-earth ions over a broad range of concentrations, wavelengths, and transitions at any temperatures.

The problems and solution methods outlined above determine the direction of research described below.

All energy migration processes can be divided into spectrally-selective and spectrally-nonselective classes. The first group of processes which represent the subject of the present section includes low-temperature spectral migration under conditions where homogeneous broadening is much lower than inhomogeneously broadening ($\delta \ll \Delta$). This condition excludes the possibility for a direct overlap of the zero-phonon lines of closely-proximate centers and is responsible for the decisive contribution to migration of nonresonant ion-ion interactions that produce phonons of energies of the order of the inhomogeneous width: $\hbar\omega \sim \Delta$ (see Section 1). The strong exponential dependence of the efficiency of this transfer process on phonon energy and, therefore, on the energy mismatch between the centers, generates a strong deformation (selectivity) of the final spectral distribution of excited and emitting centers compared to the inhomogeneous distribution in the ground state.

The spectrally-selective group of processes can also include the opposite case of energy migration due to a direct overlap of the homogeneously-broadened spectra of selectively-excited centers and centers with closely-proximate frequencies: $E_L - E = \epsilon \sim \delta$. The efficiency of this interaction is inversely proportional to the squared energy mismatch ϵ^{-2} which is also responsible for the significant selectivity (spectral selectivity) of the migration process, which is manifested as a characteristic, time-variable deformation of the spectral distribution of the excited centers.

5.1. Phonon-Assisted Spectrally-Selective Low-Temperature Migration

The Eu^{3+}, Yb^{3+}, Nd^{3+} ions represent some of the most important rare-earth ions for practical application and are convenient model systems for investigating energy migration processes. The moderate number of Stark sublevels in the ground and metastable states (between one and five) and the significant Stark and spin-orbital splitting levels which exceed the inhomogeneous width of the 1-1 resonant transition make it possible to analyze the shape of its contour (to detect spectral migration) without overlapping with other electron transitions. The analysis of interaction models carried out in Section 1, the results from studies of Stark splitting diagrams from selection of various Eu^{3+}, Yb^{3+}, and Nd^{3+} particle centers carried out in the first paper in the present volume as well as analyses of the temperature dependences of homogeneous spectral broadening compared to inhomogeneous broadening demonstrate that by varying the excitation frequency and temperature of a specimen we can effect substantial changes in the donor-donor interaction rates, models, and mechanisms as manifested in spectral-spatial migration.

Under conditions of low temperature when thermal energy is insufficient to achieve any significant transitions to the excited Stark or vibrational states ($n(\Delta E) \rightarrow 0$ for $kT \ll \Delta E$) we consider Eu^{3+} and Yb^{3+} ions as two-level systems, taking into account only inhomogeneous broadening of their energies $n(E)$ through continuously varying shifts from resonance ϵ. The differential is observed for Nd^{3+} ions for which, as shall be demonstrated below, it is necessary in many cases to account for the multilevel nature of the resonance shift even at $T = 4.2$ K.

The low-temperature also severely suppresses the direct inter-Stark one-phonon and Raman two-phonon transitions responsible for the homogeneous broadening of rare-earth ions in solids. The narrow homogeneous width which is three to five orders of magnitude narrower than the homogeneous width in turn produces the exceedingly low probability of finding direct resonance of the zero-phonon lines and permits excluding the resonant interaction mechanism from the analysis.

Analysis of the pair interaction models (see Section 1) and the schemes in Figure 1 reveals that the only possible mechanism under these conditions (T = 4.2 K) is nonresonant energy transfer with phonon emission (see Figure 1a) which does not require activation energy.

The authors of Refs. [143-145, 149] came to a similar conclusion based on experimental analyses of the stationary helium luminescence spectra of Yb^{3+} and Nd^{3+} ions and subsequently Eu^{3+} ions as well [148] in which a "red shift" (or a Stokes luminescence band) was detected under broadband, monochromatic excitation resulting from downward migration on the energy scale.[6]

An important feature of rare-earth ions in glass is the continuous set of shifts of the interacting ions ϵ; the ion-ion interaction efficiency is quite sensitive to this quantity (see Expressions (8) and (12)). On the other hand this severely complicates the theoretical analysis of the process and requires incorporating, aside from the spatial inhomogeneity, the spectral inhomogeneity of the set of centers. On the other hand, a successful resolution to this problem may make it possible to analyze the characteristic dependences of the migration rate and the ion-ion interaction efficiencies on the shift frequency or, which is the same thing, the energy of the phonon participating in the process.

Such a relation was first analyzed in Ref. [146] by a spectrally-selective measurement of the decay kinetics of centers of energy E_R on the short-wavelength wing of the inhomogeneous contour after broadband pulsed excitation. Here the authors detected Förster decay kinetics which indicated dipole-dipole Yb-Yb, Nd-Nd, Eu-Eu interaction in accordance with the theoretical results from Refs. [34, 81-83] at T = 4.2 K. Assuming excitation transfer from the centers of energy E_R runs primarily to centers at the maximum of the inhomogeneous band $E = E_{max}$ and assuming the transfer efficiency is independent of E, they determine that the relation C (ϵ) is exponential in nature (by setting $\epsilon = E_R - E_{max}$). This result has not been directly explained in energy transport theory and may be related to the insufficient accuracy of the method which has a limited range of applicability: $\Delta \ll E_R - E_{max} \ll \Delta E$. We also note the limited temperature range for using this method resulting from the transfer irreversibility condition ($kT \ll \Delta$).

The use of a monochromatic excitation source in Refs. [148, 149] to

---

[6] The interpretation of the results from Refs. [143, 145] for Nd^{3+} ions is not unique, as shall be demonstrated below.

investigate low-temperature migration eliminated a number of the constraints of the preceding technique, although the limited stationary selective spectroscopy method allowed the authors to draft only the frequency dependences of the averaged migration rate without converting to the microefficiency of ion-ion interaction $C(\epsilon)$.

Spectral migration kinetics. The primary tasks in our investigation of low-temperature energy migration include the selection and development of the most direct and valuable experimental techniques as well as the creation of theoretical models to establish a relation between observed phenomena and the elementary ion-ion interaction microprocesses.

Kinetic selective laser spectroscopic techniques were used as the experimental basis as these permit not only observation of the migration phenomenon, but also direct measurement of its kinetics and rates, all the while applicable over a broad range of temperatures and a substantial portion of the spectrum.

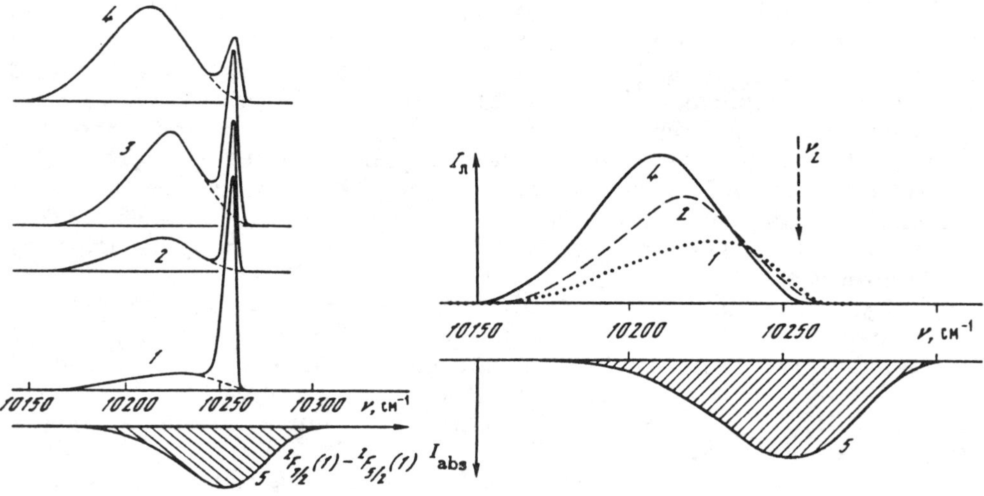

Fig. 3. Spectrally-kinetic manifestations of ion-ion interactions of Yb^{3+} ions in Ba-Al-phosphate glass at T = 4.2 K.

Spectral measurement time relative to excitation time (t_D): 1 - 50 μs, 2 - 800 μs, 3 - 1.2 ms, 4 - 2 ms, 5 - absorption spectrum.

Fig. 4. "Red" shift and change in the shape of the Stokes wing of the luminescence spectrum due to repeated low-temperature migration events plotted as a function of delay.

1, 2, 4, 5 - same as in Figure 3.

One such kinetic technique is the time-resolved method of recording luminescence spectra following pulsed monochromatic excitation. The most important advantage of this method is the capability for controlled observation of any of a variety of stages in the spectral migration process with random concentrations of active ions. Unlike the stationary spectroscopy techniques [148, 149] the scale of the observed nonstationary transport process in time-resolved spectroscopy is not determined by the average luminescence decay time $\bar{\tau}$ but by the delay t_D between the recording time and the excitation time which is set by the researcher during the experiment.

We employed this technique to analyze spectral energy migration among Yb^{3+} and Nd^{3+} ions in phosphate glasses [31, 150-153].

The Yb^{3+} *ion.* Figure 3 demonstrates the temporal evolutions of low-temperature ($T = 4.2$ K) luminescence spectra of the $^2F_{5/2}(1)$—$^2F_{7/2}(1)$ resonant transition for Yb^{3+} ions in Ba-Al-phosphate glass (concentration $n = 2.9 \cdot 10^{20}$ cm^{-3}) under abrupt selective excitation ($\Delta t_E = 10^{-8}$ s) at the center of the inhomogeneously-broadened absorption band (see Figure 3, curve 5) with $v_E = E_L = 10,256$ cm^{-1}. It is clear that the narrow resonance peak in the spectra diminishes with increasing delay due to the luminescence of the resonantly-excited centers, which also produces an increase in the broad, inhomogeneous "pedestal" for which the centers acquiring the energy from migration from the initial sites are responsible. Here energy migration will be observed only the Stokes (relative to E_L) range of the inhomogeneously-broadened band.[7]

The absence of anti-Stokes excitation transfer to centers of energies $E > E_L$ ($E - E_L > kT$) confirms the frequency-dependent, nonresonant nature of energy migration. At low temperatures ($kT \ll \Delta$) this is characterized by emission of a phonon in each ion-ion energy transfer event where the rate of this process is determined not only by the number $n(E)$ of acceptors[8] — Yb^{3+} ions — acquiring the excitation energy, but also by the density of phonon states in the medium at the resonance shift frequency of the energies of the interacting transitions $\epsilon = E_L - e$ (See Section 1). It is clear that the hole in the figure between the narrow and broad spectral components, as in the case of the intracenter vibronic transitions (see Figure 13 in the first paper of this volume) is due to the low density of phonon states in the low vibrational frequency range (0-20 cm^{-1}) in spite of the fact that the density of the electron states of the acceptors is maximized in this range.

Given the external similarity of the two processes they nonetheless differ in that one occurs in isolation at each optical center and is independent of the impurity concentration and the observation time, while the other, on the other hand, is typically an ion-ion process accompanied by spatial energy transfer whose rate grows with increasing impurity concentration and is manifested as a significant evolution of the spectrum over time.

---

[7] See footnote 1 in the first paper of the present volume.

[8] In this section the "acceptor" centers will not refer, unlike the case in Section 2, to the quencher centers which acquire energy and dissipate such energy as heat, but rather will refer to centers that acquire and reradiate electron excitation energy.

Therefore an increase in the delay will not only result in a significant redistribution of intensities between the narrow resonant spectral component and the broad Stokes spectral component but will also cause a sequential time shift of the Stokes band towards the low frequency range of the spectrum (Figure 4). This can be attributed to the repeated irreversible nonresonant energy transfer events (acceptor-acceptor) down the energy scale with phonon emission which serve to redistribute the excitation energy within the Stokes band with preferential distribution to the low-energy centers.

Analysis demonstrates that in the case of irreversible spectral energy migration among a set of like, yet spectrally-nonequivalent centers, the decay kinetics $I(E_L, t)$ of a narrow, resonant peak (the donor centers) will be accurately described by the existing Förster kinetics generalized to account for inhomogeneously broadening which is easily done by assuming a microscopic broadening mechanism (for example, disorder of the ligands for glasses) which does not correlate with the distances between impurities [27, 28, 154]:

$$I(E_L, t) = \exp[-\gamma_F (E_L) t^{3/S}], \tag{143}$$

where

$$\gamma_\Gamma (E_L) = \int_0^\infty \gamma_F (E_L, E) dE = (4/3) \pi \Gamma (1 - 3/S) n \int_0^\infty C_{DA}^{3/S} (E_L, E) dE. \tag{144}$$

Unlike Expression (31) here the macroparameter $\gamma_F(E_L)$ has an integral meaning, characterizing the sum energy transfer rate $W_F = \gamma_F^{S/3}$ from the donor-centers of energy E_L to the entire set of spectrally-nonequivalent acceptor-centers of energies $E \neq E_L$. Assuming similar luminescence decay kinetics for the donors and acceptors as well as radiative transition probabilities A which is not too rough an approximation in our case where we employ an arbitrary division[9] we can write the relation between the integral intensities of the broad acceptor band and the narrow resonance donor luminescence peak (area $I_W(E, t_D)$ and $I_N(E_L, t_D)$) with the parameter $\gamma_F(E_L)$:

$$\lg \left[\frac{I_N(E_L, t)}{I_W + I_N}\right] = -0.434 \gamma_F(E_L) t_D^{3/S}. \tag{145}$$

Figure 5 shows results from such processing of time-resolved spectra for two laser excitation frequencies (10,256 and 10,288 cm^{-1}) plotted as linear (curves 1 and 3) and exponential (power of 1/2, curves 2 and 4) time dependences ($S = 6$). The macroparameters γ_F and rates W_F) found from the

---

[9] The dispersion of the spontaneous decay probability $A(E)$ and the lifetime $T(E)$ can be accounted for by dividing the luminescence spectrum $I_L(E)$ by the correction function $EA(E)\exp[-t/T(E)]$.

Fig. 5. Spectral energy migration kinetics of Yb^{3+} ions in Ba-Al-phosphate glass at selective laser excitation frequencies $v_{E_L} = 10,256$ cm^{-1} (a) and 10,288 cm^{-1} (b).

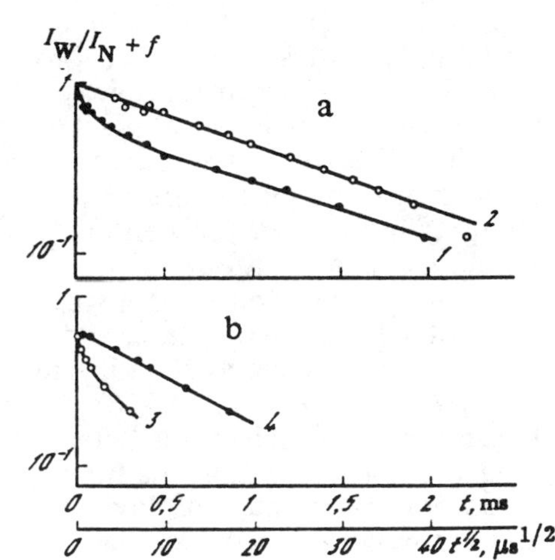

Fig. 6. Schemes for measurement of the decay kinetics of the donor centers $hv_1 = E'_R = E_L$ and the build-up kinetics of the acceptor centers $hv_2 = E''_R \neq E_L$ by the double selection method based on excitation E_L and recording E_R energies.

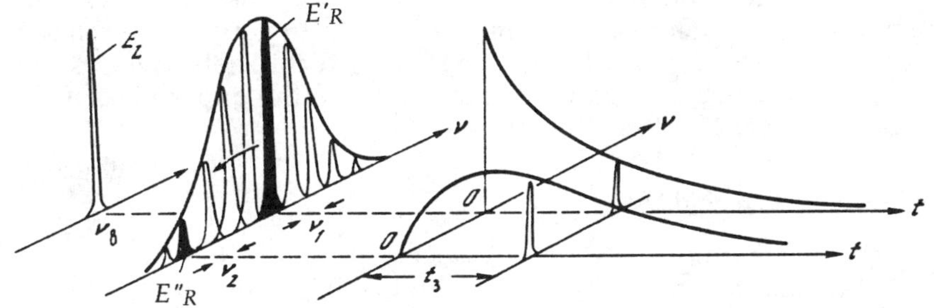

slope of curves 2 and 4 were equal to 40 s$^{-1/2}$ (1.6 • 10^3 s^{-1}) and 60 s$^{-1/2}$ (3.6 • 10^3 s^{-1}), respectively. The increase in the macroparameter $\gamma_F(E_L)$ reflecting the substantial increase in the energy migration rate with increasing donor excitation energy E_L results from the increasing number of effective energy acceptors (Yb^{3+} ions having transition energies $E < E_L$) and the increase in the number of phonons involved in nonresonant energy transfer ($\epsilon = E_L - E$).

Another kinetic method which can be used successfully to investigate spectral energy migration involves recording the resonant luminescence decay process by means of double spectral selection: selective laser excitation and simultaneous selective luminescence recording [125, 151].

Although this method is not as easily employed as the process of recording nonstationary spectra it has the advantages of providing continuous observation of the decay kinetics of the "donor" or the build-up kinetics of the "acceptor" centers in time, which simplifies data processing and eliminates the uncertainty in the normalization of the time-resolved luminescence spectra. Figure 6 shows the schemes used for double selection and measurement of the decay kinetics of the "donor" centers ($E'_R = E_L$) and the build-up kinetics of the "acceptor" ($E''_R \neq E_L$). The results from using this technique to measure the decay kinetics of donor centers $I_N(E_L)$ for Yb^{3+} ions in Ba-Al-phosphate glass at a concentration of $2.9 \cdot 10^{20}$ cm^{-3} are shown in Figure 7. The curves are measured at $T = 4.2$ K for a number of discrete groups of centers ($E_R = E_L$) of energies on both the short-wavelength wing and in the center as well as the long-wavelength wing of the inhomogeneous contour of the $^2F_{5/2}(1)-^2F_{7/2}(1)$ resonant transition. This could not have been done using the technique described in Refs. [6, 146] without selective excitation.

Figure 8 shows the kinetic curves of low-temperature energy migration $I(E_L, t) = I_N(E_L, t)/I_{sp}(E_L, t)$ calculated using the data from Figure 7 for selectively-excited and selectively-recorded luminescence kinetics. Here $I_{sp}(E_L)$ are the spontaneous luminescence kinetics of the donor centers measured for a specimen of low activator concentration ($n = 2.9 \cdot 10^{19}$ cm^{-3}) and zero energy migration, while $I_N(E_L)$ are the luminescence kinetics of a specimen of concentration $n = 2.9 \cdot 10^{20}$ cm^{-3} with nonradiative energy migration along the inhomogeneous contour. The flattening of the relation at coordinates $I(E_L) - t^{1/2}$ (over a broad range of t) for all selective excitation frequencies E_L indicates a close correspondence of the experimental decay curves to a theoretical law (see Expression (143))

$$\ln I(E_L) = -\gamma_F(E_L)\sqrt{t} \qquad (146)$$

for dipole-dipole ion-ion interaction ($S = 6$).

This fact can be taken as experimental confirmation of the theoretically predicted [27, 154] confirmation of the Förster nature of the "donor" decay kinetics under joint action of spatial and spectral disorder.

The slope of the experimental curves in Figure 8, as follows from Expression (145), yields the integral migration macroparameter $\gamma_F(E_L)$ which determines the average excitation drift rate from the initial site (donor): $W_F(E_L) = \gamma_F^2(E_L)$ (for $S = 6$). The dependences of the macroparameter γ_F and the average rate W_F on the selective excitation energy of Yb^{3+} ions are shown in Figure 9. Here the selective excitation and recording energies are plotted on the X-axis (in cm^{-1}), while the corresponding values of γ_F and W_F are plotted on the Y-axis. The inhomogeneously-broadened absorption spectrum of the resonant transition reflecting the distribution of the Yb^{3+} optical centers over the frequencies of the $^2F_{5/2}(1)-^2F_{7/2}(1)$ transition are shown in the lower half of the figure. As we see $\gamma_F(10,256) = 35$ s$^{-1/2}$, $\gamma_F(10,288) = 56$ s$^{-1/2}$ are in good agreement with analogous values of $\gamma_F(E_L)$ calculated for the same selective excitation frequencies from the time-resolved luminescence spectra. This suggests that the assumptions used in processing the spectra in Figure 3 were reasonable.

The $\gamma_F(E_L)$ relations obtained here demonstrate the sharp rise in the sum energy migration rate $W_F(E_L)$ with increasing selective excitation fre-

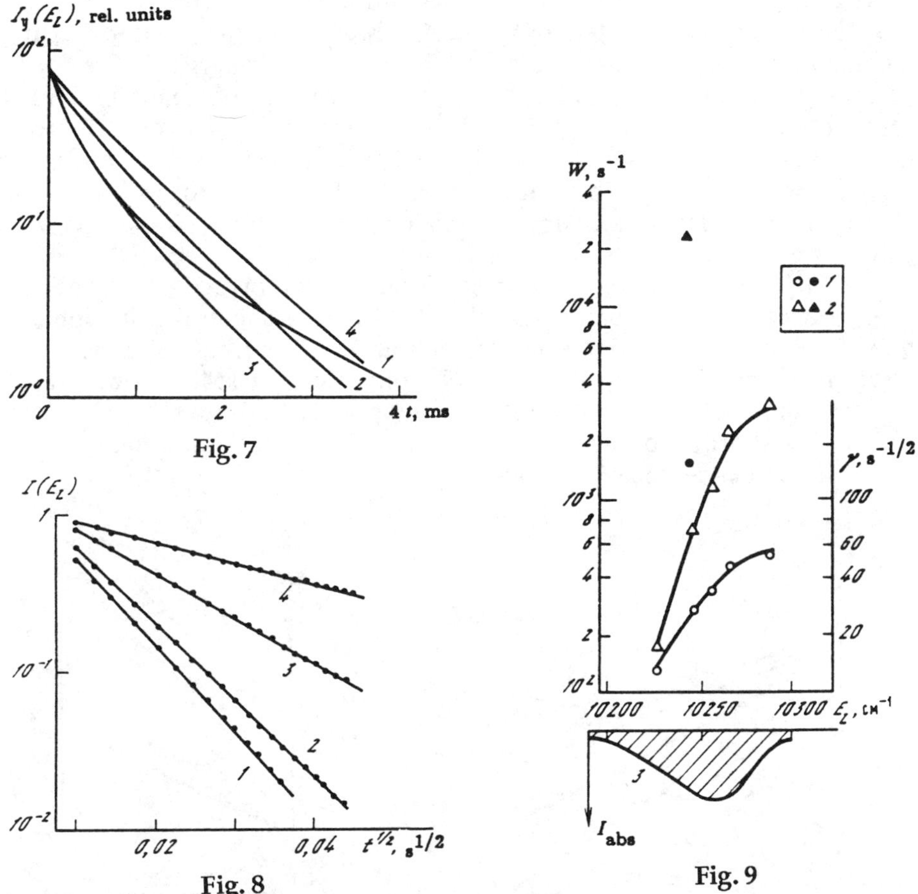

Fig. 7. Luminescence decay kinetics of Yb^{3+} ions in Ba-Al-phosphate glass $n = 2.95 \cdot 10^{20}$ cm^{-3} under selective laser excitation and selective recording within the inhomogeneous contour of the $^2F_{5/2}(1)$—$^2F_{7/2}(1)$ transition at $T = 4.2$ K.

E_L cm^{-1}: 1 - 10,288, 2 - 10,266, 3 - 10,245, 4 - 10,224.

Fig. 8. Kinetic curves of low-temperature Yb^{3+} - Yb^{3+} energy migration at I - $t^{1/2}$ calculated in accordance with Figure 7.

1-4 - same as Figure 7.

Fig. 9. Macroparameter $\gamma(1)$ and energy migration rate $W(2)$ plotted as a function of selective excitation energy of Yb^{3+} ions at the $^2F_{5/2}(1)$—$^2F_{7/2}(1)$ transition in Ba-Al-phosphate glass.

Curve 3 - inhomogeneously-broadened absorption spectrum; open symbols: $T = 4.2$ K, solid symbols: $T = 80$ K.

quency (energy) E_L. Such a significant increase in the average transfer rate for the higher frequency "donor" centers is due to the increase in the number of "acceptor" ions (centers of energies $E < E_L$) and the greater density of phonon states at frequencies corresponding to greater energy shifts: $\epsilon = E_L - E$. An increase in temperature to 80 K causes a significant rise of the macroparameter ($\gamma(10,245) = 152$ s$^{-1/2}$) and the spectral energy migration rate ($W(10,245) = 2.3 \cdot 10^4$ s^{-1}, see Figure 9) which demonstrates an increasing efficiency of Yb^{3+} - Yb^{3+} interactions due to temperature stimulation of Stokes and particularly anti-Stokes transfer (a more detailed analysis of this case is given below).

Nd^{3+} *ion.* We employed the same method of investigating the luminescence decay kinetics using the double resonant spectral selection technique ($E_R = E_L$) to investigate low-temperature (T = 4.2 K) spectral energy migration among Nd^{3+} ions in Li-La-Nd-phosphate glass [31, 150, 151]. The concentration of Nd^{3+} ions was set at a moderately high level ($n(\mathrm{Nd}^{3+})$ = 2.7 $\cdot 10^{20}$ cm^{-3}) in order to eliminate the effect of cross-relaxational quenching on the luminescence decay kinetics.

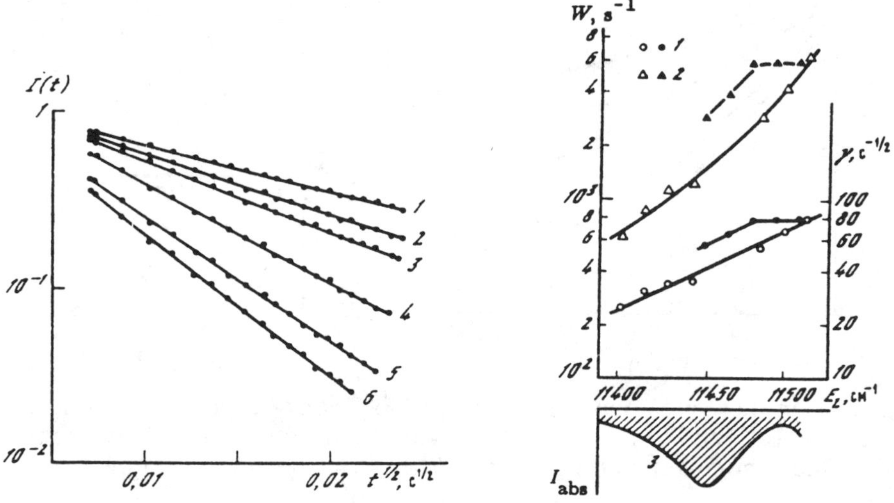

Fig. 10. Kinetic curves of low-temperature (T = 4.2 K) Nd^{3+} - Nd^{3+} energy migration at $I - t^{1/2}$ for Li-La-Nd-phosphate glass.
n = 2.7 $\cdot 10^{20}$ cm^{-3}; E_L, cm^{-1}: 1 - 11,403, 2 - 11,416, 3 - 11,429, 4 - 11,468, 5 - 11,494, 6 - 11,507.

Fig. 11. Energy migration microparameter γ (1) and its rate W (2) plotted as a function of selective excitation energy at the $^4I_{9/2}(1)-^4F_{3/2}(1)$ transition in Li-La-Nd-phosphate glass.
Curve 3 - inhomogeneously-broadened absorption spectrum; open figures: T = 4.2 K; solid figures - T = 77 K.

One feature of the spectral properties of Nd^{3+} ions in glass which complicates analysis of spectral energy migration is the complex electron structure of the ground and excited levels resulting from the low average Stark splitting levels ΔE which are comparable to the inhomogeneously-broadened level Δ which causes overlapping of the inhomogeneous lines. As a result even at $T = 4.2$ K it is necessary to take into account the multilevel nature of electron resonances in the case of interaction in the active ion pair, while the maximum selectivity of monochromatic excitation drops off precipitously with increasing temperature ($kT \sim \Delta E \approx 50$ cm^{-1}).

Analysis of the selective luminescence decay kinetics of Nd^{3+} ions using analogous techniques for the Yb^{3+} ion case described above also demonstrates the characteristic power-law (power of 1/2) relation $I(E_L, t)$ (Figure 10) which is a reflection of the dipole-dipole interaction of Nd^{3+} ions in this matrix ($S = 6$). Here the macroparameter γ_F turned out to be highly dependent on the donor excitation energy E_L (Figure 11). Part of the inhomogeneously-broadened absorption spectrum of the $^4I_{9/2}(1)-^4F_{3/2}(1)$ transition of width $\Delta = 100$ cm^{-1} is shown in the lower section of the figure.

As for Yb^{3+} ions at 4.2 K the rise of $\gamma_F(E_L)$ for Nd^{3+} ions can be attributed to two factors: the increasing concentration of centers acquiring the energy (acceptors) and the increasing microefficiency of ion-ion interaction $C(E_L - E)$ with significant resonance shifts ($E - E_L$). However, unlike the Yb^{3+} ions there are several causes of the rise in $\gamma_F(E_L)$. In the range of E_L from 11,400 to 11,470 cm^{-1} the primary mechanism includes nonresonant $Nd^{3+} - Nd^{3+}$ reactions with emission of acoustical phonons, while the enhancement of such interactions, as in the case of Yb^{3+} ions, is due to the greater density of phonon states at frequencies corresponding to the greater energy shifts: $\epsilon = E_L - E$. The contribution of the resonant $Nd^{3+} - Nd^{3+}$ interaction mechanism is enhanced with increasing frequency ($v_E = E_L > 11,470$ cm^{-1}) even at $T = 4.2$ K, since it is possible to observe the actual overlap integrals of the luminescence spectrum of the donor electron transition (at the second Stark component $^4I_{9/2}(2)$) and the absorption spectrum of the $^4I_{9/2}(1)-^4F_{3/2}(1)$ resonant transition at the "acceptors" for these groups of centers. Here, as we see from Figure 12, donor centers having increasingly larger overlap integrals with the acceptor group are excited with increasing E_L, which is manifested as an increase in the interaction efficiency.

As we see the doubling of the average migration rate W_F when the excitation frequency jumps from 11,473 to 11,503 cm^{-1} correlates well with the increase in the overlap integral of the donor luminescence spectrum and the acceptor absorption spectrum $\int I_L(v)I_{abs}(v)dv$ (by a factor of 1.8) (the area under the overlap spectrum 4, 5, Figure 12).

The sharp dependence of the low-temperature energy transfer rate $W_F(E_L) = \gamma_F^2(E_L)$ on the selective laser excitation and recording frequency (see Figures 9, 11) can be used as a method of controlling the donor–donor (see Figures 9, 11) can be used as a method of controlling the donor-donor energy migration rate in a group of like, yet inhomogeneous optical centers, which is of significant interest for detecting, identifying, and analyzing the micromechanisms in complex interaction schemes characteristic of multilevel cross-relaxing active laser ions (see, for example, the quenching schemes of the Nd^{3+}, Sm^{3+}, Er^{3+} ions). Thus the long-wavelength selective laser excitation of the lower energy optical centers for $kT \ll \Delta$ with application to Nd^{3+} ions in crystals and glasses can be used to severely limit the migration-controllable quenching rate ($\overline{W} \to 0$). This in turn makes it possible to identify the most valuable disordered donor-acceptor quenching stage even at very high active ion concentrations for which a kinetic limit condition with a strictly exponential, yet low-yield deactivation kinetics occurs under traditional conditions.

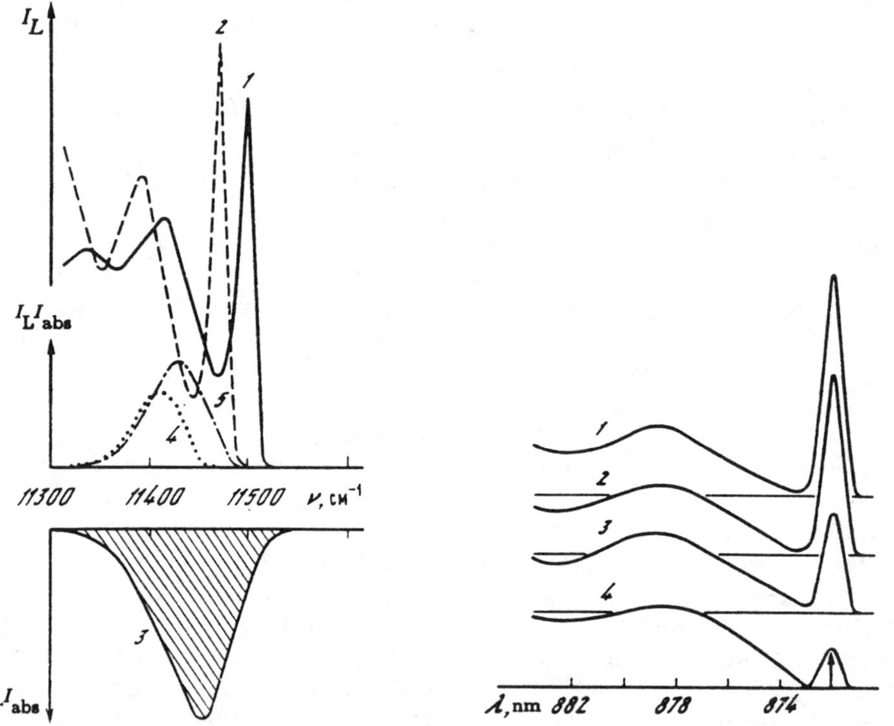

Fig. 12. Overlap of the selective luminescence (1, 3) and absorption (3) spectra of Nd^{3+} ions in Li-La-Nd-phosphate glass at $T = 4.2$ K.
1, 5 - $E_L = 11,503$ cm^{-7}, 2, 4 - $E_L = 11,473$ cm^{-1}

Fig. 13. Manifestation of low-temperature Stokes energy migration in the selective luminescence spectra of Nd^{3+} ions in Li-La-Nd-phosphate glass.
$n = 2.7 \cdot 10^{20}$ cm^{-3}, $T = 4.2$ K, $v_L = 11,468$ cm^{-1}; t_D, μs: 1 - 200, 2 - 300, 3 - 400, 4 - 1000.

The measurements carried out at $T = 77$ K show an increase in ion-ion interaction efficiency appearing in the luminescence decay kinetics of selectively-excited $I(E_L)$ centers as a more clearly-expressed nonexponential nature, and consequently, as an increase in the dependence of $\gamma(E_L)$ on temperature T. Analysis of the decay kinetics reveals that this too corresponds to the dipole-dipole Nd^{3+} interaction mechanism (having a power law (power 1/2) form) and made it possible to determine the macroparameter $\gamma(E_L)$ shown in Figure 11. It is clear from the figure that $\gamma(E_L)$ grows at $T = 77$ K in the low frequency range with increasing excitation energy E_L (and when $T = 4.2$ K), while at an E_L corresponding to 11,470 cm^{-1} an inflection is observed and the macroparameter then remains unchanged. The near two-fold temperature rise in the migration rate $W(E_L)$ at $T = 77$ K appears only in the low excitation frequency range and can be attributed to the temperature stimulation of anti-Stokes nonresonant interactions as well as the resonant interactions involving components of the $^4F_{3/2}(2)$, $^4I_{9/2}(2, 3, 4)$ levels which are partially populated at 77 K.

A different situation occurs for the short-wavelength centers. The interaction for these centers is always Stokes interaction and can be resonance interaction as early as $T = 4.2$ K (see Figure 12). The temperature rise in this case only redistributes the contributions from the various electron resonances to the sum interaction efficiency which, due to the similar oscillator strengths of the inter-Stark transitions, will result in a weak temperature dependence of the migration rate for the short-wavelength centers.

In addition to kinetic measurements we also carried out spectral measurements of nonstationary luminescence which provide a more clear picture of the spectral energy migration dynamics.

The luminescence spectra were recorded at the $^4F_{3/2}$–$^4I_{9/2}$ transition with a time delay for a specimen with $n(Nd^{3+}) = 2.7 \cdot 10^{20}$ cm^{-3} by means of laser pumping to the long-wavelength component of the absorption spectrum of the $^4I_{9/2}(1)$–$^4F_{3/2}(1)$ transition for $T = 4.2$ K which provides good excitation selectivity. It follows from Figure 13 that, as in the case of Yb^{3+} ions, energy migration among the Nd^{3+} ions runs only towards the Stokes (the long-wavelength) portion of the spectrum. The position of the long-wavelength pedestal which grows in time (i.e., the spectral position of the acceptor centers) nearly coincides with the spectral contour of the $^4F_{3/2}(1)$, $^4I_{9/2}(2)$ transition at the second Stark component of the ground state. Area normalization of the luminescence spectra of $^4F_{3/2}(1)$–$^4I_{9/2}(1, 2)$ (as was carried out for the Yb^{3+} ions) makes it possible to determine the integral decay kinetics of the donors $I(E_L, t)$ (Figure 14). These kinetics, similar to those in Figure 10, are accurately described by a Förster law with an average rate $W_F = 2.8 \cdot 10^3$ s^{-1}, corresponding to the data in Figure 12.

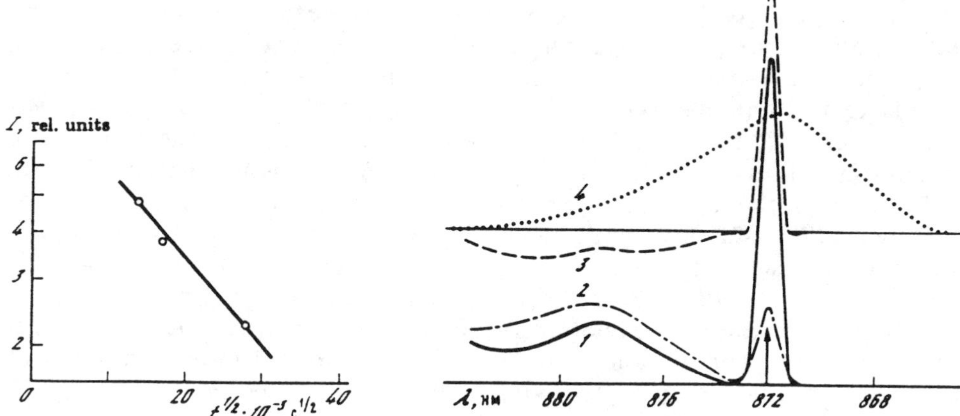

Fig. 14. Low-temperature Nd^{3+} - Nd^{3+} energy migration kinetics based on data from Figure 13.

Fig. 15. Identification of the acceptor luminescence spectrum of Nd^{3+} ions and a comparison to the energy distribution function of the centers (absorption contour of $^4I_{9/2}(1)-^4I_{3/2}(1)$).

Analysis of the spectral distribution of the "acceptor" luminescence reveals the efficiency of the transition between Nd^{3+} ions with different resonance shifts $\epsilon = E_D - E_A$ within the inhomogeneously-broadened contour. We plotted the difference spectrum (Figure 15, curve 3) of the normalized donor luminescence contour ($t_L = 20$ μs $\rightarrow$ 0, curve 1) and the acceptance luminescence contour ($t_D = 800$ μs $\rightarrow$ ∞, curve 2) at the $^4F_{3/2}-^4I_{9/2}$ transition for this purpose. The narrow peak in curve 3 shows the absolute decay of donor luminescence, while the broad band shows the growth of acceptor luminescence. The significant difference in the shapes of the acceptor luminescence band and the energy distribution function of the centers (absorption contour of $^4I_{9/2}(1)-^4F_{3/2}(1)$, see Figure 15, curve 4) indicates the essentially selective nature of low-temperature migration. We note that the long-wavelength shift in the luminescence spectrum of Nd^{3+} during energy migration, similar to the Yb^{3+} case (see Figure 3) is not an unambiguous criterion of the resonant interaction mechanism for Nd^{3+} ions. As we see from Figures 12 and 15 it can also occur with spectrally-selective resonant transfer involving the second and third upper Stark components of the

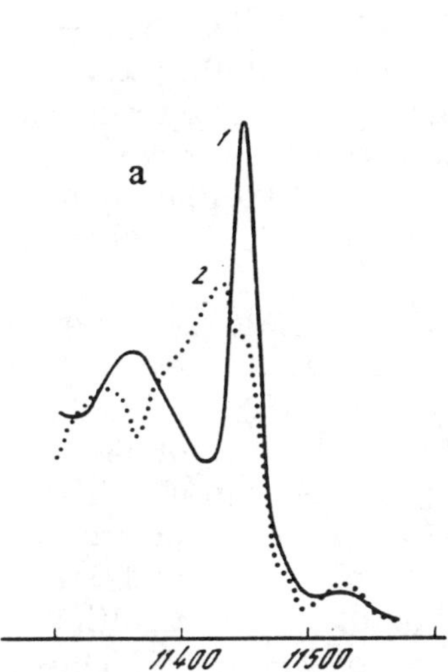

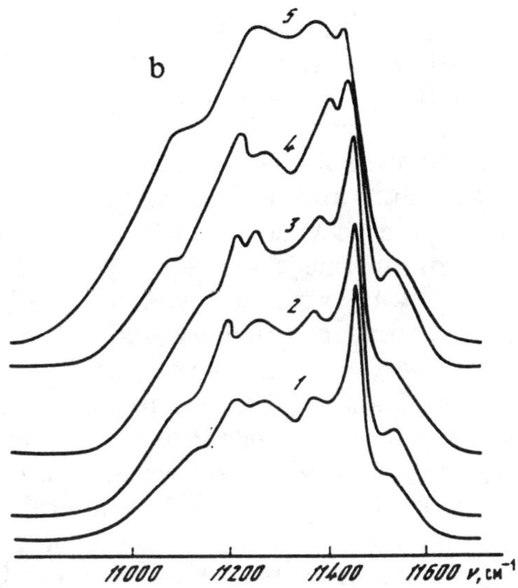

Fig. 16. Spectral Nd^{3+} - Nd^{3+} energy migration in the nonstationary selective luminescence spectrum of Nd^{3+} ions in Li-La-Nd-phosphate glass.

($n = 2.7 \cdot 10^{20}$ cm^{-3}) for $T = 77$K, $E_L = 11,446$ cm^{-1} (a) and $T = 138$ K, $E_L = 11,441$ cm^{-1} (b) for different delays t_D.
a: 1 - 40, 2 - 400 μs; b: 1 - 20, 2 - 100, 3 - 200, 4 - 300, 5 - 400 μs.

Fig. 17. Variations in the spectral distributions of excited Nd^{3+} ions manifested in the normalized luminescence (1, 4, 5) and difference (2, 3) spectra from spectral migration.

Spectra 1, 4, 5 are obtained by normalization of the corresponding spectra in Figure 16b; 2 - $I_1 - I_s$; 3 - $I_4 - I_s$.

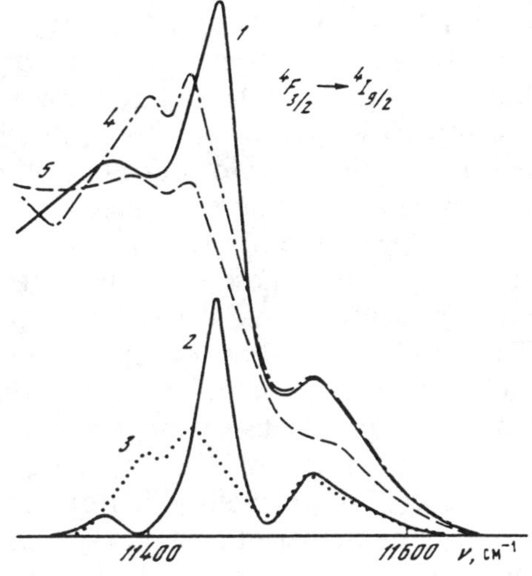

$^4I_{9/2}$ level. However, the characteristic two-hump shape of the acceptor luminescence spectrum indicates the very probable participation of acoustical phonons in this process.

Analysis of the high-temperature ($T = 77$ and 136 K) nonstationary selective luminescence spectra of Nd^{3+} ions in Li-La-Nd-phosphate glass shown in Figures 16, 17, demonstrates the substantial selectivity of the spectral migration process, as is the case for Eu^{3+} and Yb^{3+} ions. However, in addition to the similar resonant and reversible nature of the excitation migration process, a significant spectral asymmetry (in the low-frequency direction) of high-temperature migration has been identified for Nd^{3+} ions, unlike Eu^{3+} and Yb^{3+} ions; this migration can be attributed to the asymmetry of the spectral overlap at the interacting multi-Stark luminescent and absorption transitions (see Figure 12).

Analysis of the spectra in Figures 16b and 17 permit estimation of the spectral-spatial energy migration rate among the metastable levels of the Nd^{3+} ion for $n = 2.7 \cdot 10^{20}$ cm^{-3} and $T = 136$ K: $W \approx 2 \cdot 10^4$ s^{-1} [154]. This in turn makes it possible to calculate the energy transfer rate for the most important laser concentrations of Nd^{3+} ions in highly-concentrated Li-La-Nd phosphate glass. As we have noted previously the average spectral migration rate in dipole-dipole interaction is described by $W = \gamma^2$ and is proportional to the squared acceptor concentration in the glass, resulting in substantial migration rates even at a low temperature of $T = 138$ K. Thus at $n = 8 \cdot 10^{20}$ cm^{-3} the rate $W = 1.8 \cdot 10^5$ s^{-1}, and when $n = 2.7 \cdot 10^{21}$ cm^{-3} it equals $2 \cdot 10^6$ s^{-1}. The high energy exchange rates between the various Nd^{3+} optical centers in a spectrally inhomogeneous laser medium and the strong homogeneous broadening for these glasses suggests a quasidonor nature of the spectroscopic properties of this laser material, at least in the millisecond and microsecond duration range which must be accounted for when designing lasers employing such media.

Frequency dependence of interaction microefficiency. The analysis shows that the most important feature of the observed low-temperature energy transport between like impurities is its spectrally-selective nature. This is manifested as a significant dependence of the energy transfer efficiency from "donor" to "acceptor" on their relative frequencies. At the same time the continuous set of "acceptor" centers with different electron transition frequencies in the inhomogeneous medium makes this a complex, multiacceptor problem. This excludes the possibility of using common (in the case of identical acceptors) Expression (31) to find the ion-ion interaction microparameter C_{DA} from the macroparameter γ.

We proposed a method of determining the ion-ion interaction microparameters C_{DA} and their frequency dependences $C_{DA}(E_L, E)$ in Ref. [151] based on an investigation of the nonstationary luminescence build-up spectra of the Stokes band (the acceptors) (see Figure 4).

By analogy to the one-acceptor problem [1] in the case of irreversible energy transfer from the excited donors (n_D^*) to the disordered acceptor set the kinetic equations for their populations can be given as

$$\frac{dn_{\mathrm{D}}^*}{dt} = -\frac{n_{\mathrm{D}}^*}{\tau_{\mathrm{D}}} - n_{\mathrm{D}}^* \sum_i \frac{3}{S} \gamma_i(E_L) t^{3/S-1},$$

$$\frac{dn_{\mathrm{A}_1}^*}{dt} = -\frac{n_{\mathrm{A}_1}^*}{\tau_{\mathrm{A}}} + n_{\mathrm{D}}^* \frac{3}{S} \gamma_1(E_L) t^{3/S-1}, \tag{147}$$

$$\cdots\cdots\cdots\cdots\cdots\cdots\cdots\cdots\cdots\cdots\cdots$$

$$\frac{dn_{\mathrm{A}_i}^*}{dt} = -\frac{n_{\mathrm{A}_i}^*}{\tau_{\mathrm{A}}} + n_{\mathrm{D}}^* \frac{3}{S} \gamma_i(E_L) t^{3/S-1},$$

where the index i labels different types of acceptors.

In the case of lifetime equality ($\tau_{\mathrm{D}} = \tau_{\mathrm{A}} = \tau$) for the sum donor and acceptor population we have

$$dn_{\mathrm{D}}^*/dt + \sum_i dn_{\mathrm{A}_i}^*/dt = (n_{\mathrm{D}}^* + \sum_i n_{\mathrm{A}_i}^*)\tau^{-1}, \tag{148}$$

$$n_{\mathrm{D}}^*(t) + \sum n_{\mathrm{A}_i}^*(t) = [n_{\mathrm{D}}^*(0) + \sum_i n_{\mathrm{A}_i}^*(0)]\exp(-t/\tau). \tag{149}$$

Substituting the known form of donor decay kinetics into the last expression (see Equation (146) for n_{D}^*))

$$n_{\mathrm{D}}^*(t) = n_{\mathrm{D}}^*(0)\exp\left(-t/\tau - \sum_i \gamma_i t^{3/S}\right), \tag{150}$$

we find the luminescence law for a random acceptor:

$$n_{\mathrm{A}_i}^* = \frac{\gamma_i}{\sum_i \gamma_i} n_{\mathrm{D}}^*(0)\left[1 - \exp\left(-\frac{t}{\tau} - \sum_i \gamma_i t^{3/S}\right)\right]\exp\left(-\frac{t}{\tau}\right), \tag{151}$$

where

$$\gamma_i = (4/3)\pi n_{\mathrm{A}_i} C_{\mathrm{DA}}^{3/S}\Gamma(1 - 3/S). \tag{152}$$

For the case of the continuous energy distribution of like centers we go from summation to integration, labeling the donor centers by the energy E_L and the acceptor centers E and substituting γ_i and n_{A_i} by $\gamma_{\mathrm{F}}(E, E_L)$ and $n(E)$:

$$n^*(E, E_L, t) = \frac{\gamma_{\mathrm{F}}(E, E_L)}{\gamma_{\mathrm{F}}(E_L)} n^*(E_L, 0)\left[1 - \exp\left(-\frac{t}{\tau} - \gamma_{\mathrm{F}}(E_L)t^{3/S}\right)\right]\exp\left(-\frac{t}{\tau}\right). \tag{153}$$

where

$$\gamma_{\mathrm{F}}(E, E_L) = (4/3)\pi n(E) C_{\mathrm{DA}}^{3/S}(E, E_L)\Gamma(1 - 3/S);$$

$$\gamma_{\mathrm{F}}(E_L) = \int_0^\infty \gamma_{\mathrm{F}}(E, E_L)\,dE; \qquad n = \int_0^\infty n(E)\,dE. \tag{154}$$

134   *Alimov et al.*

We utilize a single hop model in this analysis, i.e., we have assumed that repeated acceptor-acceptor energy transfer events that, as we have seen from Figures 4, 5, distort the acceptor spectrum, are absent. It is clear from Expression (153) that in this model the build-up of the various acceptors is determined by the integral decay kinetics of the donors and is independent of the acceptor electron transition energy E. However, the number of excitations delivered to the centers (the acceptors) of certain energy E (or, the Stokes band spectrum) is directly related to the energy distribution of the macroparameter $\gamma_F(E, E_L)$ which in turn is directly related to the energy distribution of the centers E and the desired frequency distribution of the ion-ion interaction efficiency microparameter $C_{DA}(E, E_L)$.

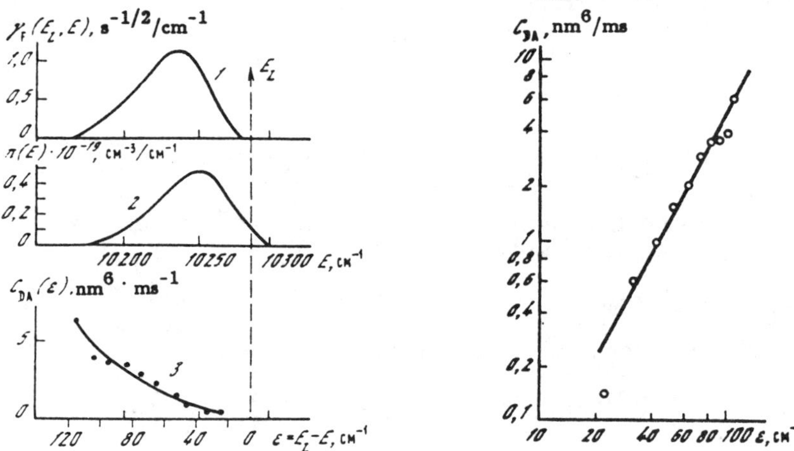

Fig. 18. Results from processing of experimental luminescence spectra of Yb^{3+} ions in Ba-Al-phosphate glass.

1 - acceptor luminescence spectrum $I(E_L, E) \sim \gamma_F(E_L, E)$, normalized to $\gamma_F(E_L)$; 2 - inhomogeneous energy distribution of centers $n(E)$; 3 - dependence of the interaction microparameters C_{DA} on the energy of the phonon assisting in the energy transfer process $\epsilon = E_L - E$.

Fig. 19. Nonresonant ion-ion interaction efficiency $C_{DA}(Yb^{3+} - Yb^{3+})$ plotted as a function of the frequency of the phonon assisting in the energy transfer process ϵ.

The results from processing the experimental spectra are shown in Figure 18. Curve 1 shows the spectrum of the Stokes "acceptor" luminescence band of Yb^{3+} ions in Ba-Al-phosphate glass ($n = 2.9 \cdot 10^{20}$ cm^{-3}) at $T = 4.2$ K 50 μs after pulsed laser excitation of energy $E_L = 10,288$ cm^{-1}. The spectrum of the Stokes luminescence band measured in zero energy migration for $n \to 0$ and $t \to 0$ was subtracted from the experimental spectrum to determine this spectrum (see Figure 11b, spectrum 2, from the first paper in this volume).

An advantage of the proposed analysis method based on a measurement of the nonstationary luminescence spectrum is the direct possibility of controlling and maintaining satisfaction of the zero repeat transfer event condition (the one hop approximation) which is achieved by using a small delay t_D in the recording process (t_D = 50 μs). Normalizing the area under the luminescence curve with E_L = 10,288 cm^{-1} and t_D = 50 μs to the integral energy transfer macroparameter for energy transfer from the donor $\gamma_F(E_L)$ = 56 s$^{-1/2}$ based on kinetic measurements (see Figure 9) we go from the luminescence spectrum to the energy distribution of the differential macroparameter $\gamma_F(E, E_L)$ (see Figure 18, curve 1). Curve 2 in Figure 18 shows the transition energy distribution of the centers $n(E)$ normalized to the total Yb^{3+} ion concentration in the glass: n = 2.9 $\cdot$ 10^{20} cm^{-3}. Dividing $\gamma_F(E, E_L)$ by curve $n(E)$ yields the spectral dependence of $(4/3)\pi\Gamma(1 - 3/S)C_{DA}^{3/S}(E, E_L)$ which we use together with the multipolarity of the interaction determined from kinetic measurements (for Yb^{3+} - Yb^{3+} S = 6 (dipole-dipole)), to find the desired dependence of the microefficiency of ion–ion interaction on the energy of the phonon contributing to the transfer process $C_{DA}(E_L - E)$ (see Figure 18, curve 3).

This relation is shown in Figure 19 on a logarithmic plot. Its sharp square-law growth with the resonance shift frequency between donor and acceptor ($\epsilon = E_L - E$) coincides with the function $g(\epsilon)$ (see Figure 13 from the first paper in this volume) reflecting the density of states of the phonons participating in the intracenter vibronic transitions. This suggests a certain contribution of the low-frequency acoustical phonons to low-temperature nonresonant energy transfer. A correlation between $C_{DA}(\epsilon)$ and $g(\epsilon)$ indicates a common density function of the vibrational states of the glass which is responsible for both the nonresonant ion–ion spectral migration mechanism and the vibronic spectrum of the individual ions which in the general case will not always hold. Analysis of the $C_{DA}(\epsilon)$ and $g(\epsilon)$ relations from the viewpoint of identifying the diagonal or nondiagonal vibronic interaction mechanism (see Expressions (8), (12), (13)) is difficult at present due to the insufficient accuracy of spectral measurements and the complexity of fully accounting for the dispersions of the radiative probabilities and electron-phonon coupling constants of the various centers.

5.2. Spectral Selective Resonant Energy Migration

One of the most popular forms of interaction of like ions at high-temperatures is resonant interaction which, as demonstrated in Section 1, is spectrally-selective for nonequivalent centers (see Expression (22)): $C_{DD} \sim 2\delta/[(2\delta)^2 + \epsilon^2]$, i.e., it is highly dependent on the resonance shift frequency.

It follows from Expression (22) that spectral energy transfer will occur with maximum efficiency to only the spectrally-proximate centers for which the shift is small or of the order of the homogeneous spectral width ($\epsilon \leqslant \gamma$) at the same time that transfer to spectrally remote centers ($\epsilon \gg \delta$) is substantially reduced: $C_{DD} \sim \epsilon^{-2}$.

We were the first to experimentally detect and investigate such spectral transfer in Ref. [108] by investigating energy migration in Eu^{3+} ions in Na-Yb-silicate glass ($n(Eu^{3+}) = 10^{21}$ cm^{-3}). In this case of spectrally-selective transfer the most convenient and valuable technique was to analyze the spectral radiation contour with different spectral recording delays. The most convenient and important element was to investigate the time dependence of the fundamental spectral parameter which was the emission line half-width; this was carried out in Ref. [108] (Figure 20). Linear time dependence of the width Δ_N of the narrow spectral component of the $^5D_0-^7F_0$ transition of Eu^{3+} in glass at $T = 300$ K over the range $t_D = 2 \cdot 10^{-5}-10^{-2}$ s can be treated within the framework of approximate solutions of an averaged balance equation.

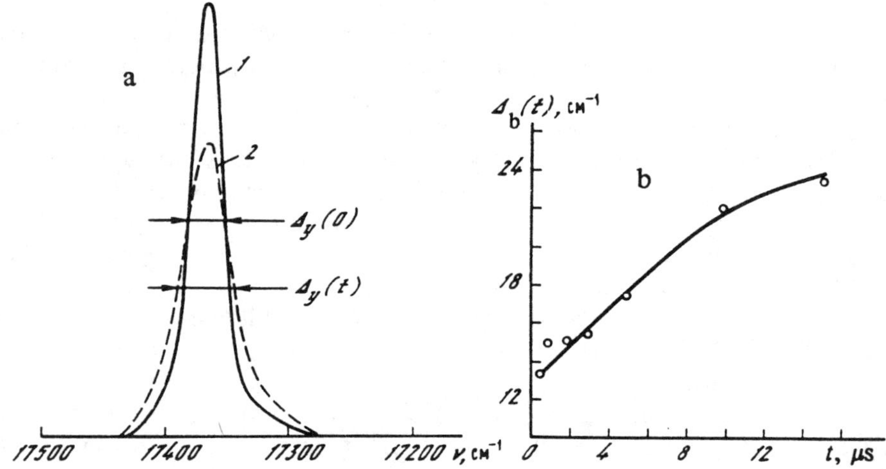

Fig. 20. Time evolution of the selective luminescence spectra of the $^5D_0-^7F_0$ transition (a) and the half-widths of the narrow resonant peak of the Eu^{3+} ion (b) in Na-Y-B-silicate glass following abrupt selective laser excitation $\nu_L = 17,361$ cm^{-1} ($n = 10^{21}$ cm^{-3}, $T = 300$ K).
a: 1 - t_D = 200 µs, 2 - 15 µs.

A theoretical analysis of such a spectral migration process in a balance approximation was carried out in Ref. [155]. The authors were the first to use the general expression described in Ref. [106] for the time dependence of the spectrally-nonuniform population in a set of interacting inhomogeneous centers

$$dP(E, t)/dt = \int [g'(E) \widetilde{W}(E - E') P(E', t) - g'(E') \widetilde{W}(E' - E) P(E, t)] dE' \tag{155}$$

and to investigate the temporal evolution of the spectral luminescence line contour of an inhomogeneous set of ions after their selective excitation. They examine three types of equilibrium energy distributions of the centers (three types of inhomogeneous broadening): rectangular, Gaussian, and Lorentzian, and accounted for two types of spectral dependences of the ion-ion interactions: the Lorentzian relation

$$\widetilde{W}_{DD}(E - E') = \frac{\widetilde{W}^0}{[(E - E')^2 + (2\delta)^2]\pi\delta^{-1}} \tag{156}$$

and an interaction that varies more substantially with energy: a rectangular relation (of width δ and height $1/\delta$)

$$\widetilde{W}(E) = \widetilde{W}^0 f(E). \tag{157}$$

For the rectangular (uniform) inhomogeneous contour

$$g'(\epsilon) = \begin{cases} 1/\Delta, & |\epsilon| \leqslant \Delta/2, \\ 0, & |\epsilon| > \Delta/2, \end{cases} \tag{158}$$

where $\epsilon = E - E'$ and following the boundary points ($|\epsilon| \ll \Delta/2$) Equation (155) is solved analytically after carrying out a Fourier transform. The solution takes the form

$$P(E, t) = \exp\left(- \frac{\widetilde{W}^0 t}{\Delta}\right) \sum_{n=0}^{\infty} \frac{(\widetilde{W}^0 t/\Delta)^n}{n!} \frac{[\Delta_y(0) + n\delta]/\pi}{(E - E_0)^2 + [\Delta_y(0) + n\delta]^2}. \tag{159}$$

Here $\Delta_N(0)$ corresponds to the width of the initial (Lorentzian) emission profile ($t = 0$) immediately after selective excitation. The sum in (159) has an extremum for $n = \widetilde{W}^0 t/\Delta$ that increases in magnitude with increasing n. In (159) the nth term can be interpreted as n sequential energy transfers, each of which shifts the edge of the emission line by 2δ (the effective spectral transfer length (see Expression (156)).

Neglecting the edge effects for $|\epsilon| < \Delta$ imposes a limit on the applicability of solution (159): $n\delta \ll \Delta$, i.e.,

$$\widetilde{W}^0 t\delta \ll \Delta^2, \quad \text{or } t \ll (\Delta/\delta)^2(\widetilde{W}^0/\delta)^{-1}. \tag{160}$$

Here the boundary time is the transfer time from the center to the edge of the rectangular profile $g'(\epsilon)$. As we see from Expression (159) the emission contour when spectral migration is present remains Lorentzian with a maximum at $E = E_0$ and a half-width $\Delta_N(0) + n\delta$. Using the expression $n = \tilde{W}^0 t / \Delta$ it is possible to obtain the time dependence of the spectral half-width $\Delta_N(t)$ as

$$\Delta_N(t) = \Delta_N(0) + (\tilde{W}^0 \delta / \Delta)t. \qquad (161)$$

As we see this linear relation is identical to our experimental observations (see Figure 20). It is possible to determine the effective spectrally-selective migration rate from this expression by knowing the characteristic spectral widths of Eu^{3+} in glass (the initial (as $t \to \infty$) spectral width $\Delta_N(0) = 13$ cm^{-1} while the homogeneous spectral width $\delta = 4.5$ cm^{-1} and the inhomogeneous spectral width $\Delta = 100$ cm^{-1}): $\tilde{W}^0 = 8 \cdot 10^4$ cm^{-1}/s.

Unlike $\tilde{W}(E - E')$ the parameter $\tilde{W}^0$ is no longer dependent on the spectral coordinates, although it includes the as yet unknown concentration dependence $\tilde{W}^0(c)$ determined by the number and type of the spatial distribution of optically-active ions in the medium. A theoretical analysis of this concentration relation $\tilde{W}^0(c)$ has not yet been carried out.

An interesting feature of Solution (159) is that when $\Delta_N(t) \gg \delta$ the time taken for spectral migration to width Δ_N in n hops is equal to (accurate to the coefficient 3) the migration time over the same spectral interval in one hop.

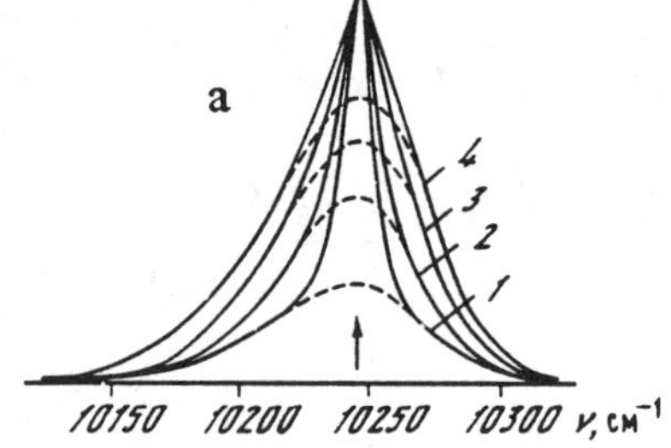

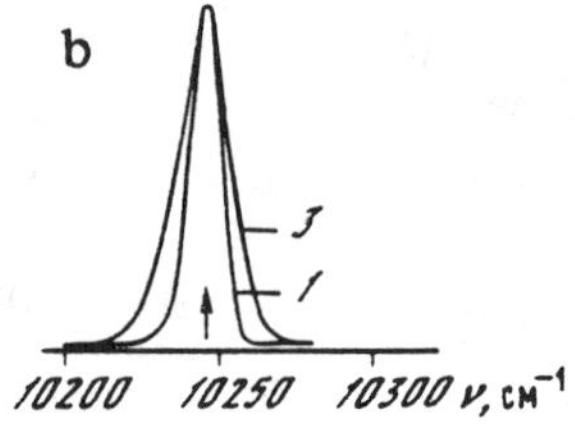

Fig. 21. Spectral migration dynamics at the $^2F_{5/2}(1)-^2F_{7/2}(1)$ transition of the Yb^{3+} ion in Ba-Al-phosphate glass ($n = 2.95 \cdot 10^{20}$ cm^{-3}, $T = 154$ K).

t_D, μs: 1 - 2, 2 - 15, 3 - 70, 4 - 100

The spectral migration calculation carried out by the authors of Ref. [155] for a rectangular dependence of interaction on the resonance shift which is a sharper dependence compared to a Lorentzian dependence

$$\widetilde{W}(\epsilon) = \begin{cases} \widetilde{W}^0/\delta & \text{for } |\epsilon| \leqslant \delta/2, \\ 0 & \text{for } |\epsilon| > \delta/2, \end{cases} \tag{162}$$

yields the spectral Gaussian relation:

$$P_n(\epsilon) = \exp\left(-\frac{\epsilon}{n\delta^2/6}\right), \tag{163}$$

whose width after n hops is proportional to $\delta\sqrt{n}$.

Accounting for the fact that each step requires identical time we see that in this case the spectral width varies as $\Delta_N \sim \sqrt{t}$. Thus similar to the diffusion case the result is a direct consequence of the severe relation $\widetilde{W}(\epsilon)$ although it seems possible that it may also be observed in the continuous Lorentzian interactions for sharply decaying energy distributions of the centers $g'(E)$ (Gaussian, Lorentzian) near the boundary conditions $\Delta_N(t) \sim \Delta$, where the precipitous weakening of interaction at width δ may not result from the specific form of $\widetilde{W}(\epsilon)$ (see Expression (162)) but rather due to the sudden drop in the density of states of the centers $g'(E)$. We observed such a weakening of the linear dependence (the transition from $\Delta_N \sim t$ to $\Delta_N \sim \sqrt{t}$) experimentally with significant delays: $t_D = 10\text{-}15$ ms where the width of the resonant peak was significant (see Figure 20), yet not sufficient to neglect the variation of $g'(E)$ in its vicinity.

An analysis of an analogous spectrally-selected resonance migration process among Yb^{3+} ions in Ba-Al-phosphate glass was carried out at a concentration $n(Yb^{3+}) = 2.95 \cdot 10^{20}$ cm^{-3} three times smaller than in the case of Eu^{3+} ions.

The nonstationary luminescence spectra at the $^2F_{5/2}(1)- {}^2F_{7/2}(1)$ transition shown in Figure 21 were measured at $T = 154$ K over a delay range from $2 \cdot 10^{-6}$ to 10^{-4} seconds. Here the arrow indicates the selective laser excitation frequency of 10,247 cm^{-1}, while the dashed line represents the spectrum of the inhomogeneously-broadened pedestal resulting from nonselective spectral migration, which will be discussed below (see Section 5.3), from the spectrum of the narrow resonant component (examples are given in Figure 21b) whose broadening due to selective migration is of interest to us [151].

Analysis of the measured spectrum made it possible for use to find the effective spectral diffusion rate $\widetilde{W} = 10^6$ cm^{-1}/s as we know the homogeneous broadening of the Yb^{3+} ion $\delta = 5.5$ cm^{-1} (near the value for Eu^{3+}) as well as the inhomogeneous broadening $\Delta(Yb^{3+}) = 55$ cm^{-1} (approximately half that of Eu^{3+}). As we see the former value even with a low particle half that of Eu^{3+}). As we see the former value even with a low particle concentration exceeds the value of $\tilde{W}^0$ by nearly an order of magnitude for the Eu^{3+} ions. This indicates a substantially greater probability of elementary resonant interactions of the Yb^{3+} ions compared to interactions of the Eu^{3+} ions: $C_{DD}(Yb^{3+} - Yb^{3+}) \gg C_{DD}(Eu^{3+} - Eu^{3+})$ which accurately correlates with the high probability of the $^2F_{5/2}(1)-^2F_{7/2}(1)$ radiative transition for the Yb^{3+} ions where migration occurs compared to the $^5D_0-^7F_0$ transition for Eu^{3+} ions.

5.3. Experimental Investigation of High-Temperature Spectral Energy Migration in a Disordered Vitreous Medium Containing Impurities

Analysis of the elementary interaction mechanisms in a pair of active ions (see Section 1) demonstrates a strong stimulating action of temperature on the energy migration process in solids. The development of energy transfer processes with absorption and stimulated emission of thermal vibrations (phonons) of the host, the thermal population of the excited Stark sublevels which incorporate new electron resonance shift together with temperature broadening of the spectral lines causing the overlap integrals of the interacting ions to rise will all normally result in higher efficiency, and a substantially more complex spectral energy migration micromechanism while weakening the frequency dependence.

The dependence of the transfer probability on the resonance shift frequency of the interacting ions ($E - E' = \epsilon$) will vanish altogether for a number of energy transfer schemes in the high-temperature range.

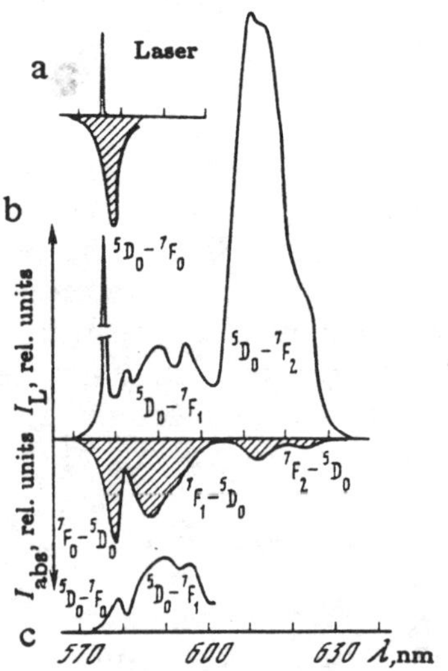

Fig. 22. Scheme of total selectivity losses in the luminescence spectrum of Eu^{3+} in Na–Y–B–silicate glass.

a - selective laser excitation scheme of Eu^{3+} at the $^7F_0-^5D_0$ transition; b - overlap of the selective luminescence spectrum and the integral absorption spectrum for $Eu^{3+} - Eu^{3+}$ interaction; c - the resulting nonselective luminescence spectrum following a single energy transfer event.

This is because the interaction begins to include a growing number of new electron or vibronic resonances in the thermal population of states somewhat above the ground state or the metastable level.

One feature of these new resonances is the significant homogeneous width, a high degree of overlapping of their luminescence and absorption spectra as well as a low degree of energy correlation of the transitions responsible for the interaction with the energy of the transition where the luminescence is recorded. This serves to equalize the transfer rates from the selectively-excited group of centers of energy E_L to any other centers of energy E within the spectrally inhomogeneous contour under analysis.

For example, at $T = 300$ K a wide range of high-temperature resonances make the primary contribution to energy transfer between the Eu^{3+} ions. Figure 22 clearly reveals strong band overlapping with a complex multi-Stark structure to the luminescence and absorption spectra: $^5D_0-^7F_2(1\text{-}5)$ and $^7F_2(1\text{-}5)-^5D_0$ as well as $^5D_0-^7F_1(1\text{-}3)$ and $^7F_1(1\text{-}3)-^5D_0$. The homogeneous width is several times greater for these transitions compared to the $^5D_0-^7F_0$ transition where spectral migration is observed. Moreover, their energy poorly correlates with the energy of the $^5D_0-^7F_0$ transition (see the first paper in this volume).

Such energy transfer schemes result in a complete loss of excitation selectivity as early as a time period characteristic of a single energy migration event. This appears in the spectrum as a broad, nonselective pedestal whose shape is similar to an inhomogeneously-broadened contour and is located under the narrow spectral component responsible for the luminescence of the selectively-excited optical centers.

The lack of spectral selectivity of the energy transfer process in no way suggests that laser excitation methods are not useful for analyzing such a process; this technique is likely to be the only method that will be successful under such conditions.

The frequency-independent nature of "high-temperature" spectral migration can substantially simplify the analysis of the luminescence spectra since in this case they decay into two spectra of variable amplitude yet with a shape that remains unchanged over time: a narrow peak and an inhomogeneous pedestal. This also makes it possible to describe the migration process by using only one value of the microparameter C_{DD} which is common to all optical centers. On the other hand this results in a substantial simplification of the theoretical analysis for spectral migration, although, on the other hand, it requires completely accounting for the reversibility of the "donor-donor" energy transfer process.

As discussed in Section 3 under these conditions there are three primary stages and two transition ranges near the boundary times t_1 and t_2 in the spectral energy migration kinetics in a disordered medium. In the early stage of "donor" luminescence decay ($t < t_1$) we have an exponential decay reflecting the order of the nearest neighbor ions with decay rates of the ensemble precluding the need to account for the reversibility of the energy transfer process and given by Expressions (48)-(50). For $t_1 < t < t_2$ this exponential decay is replaced by the characteristic nonexponential decay law which is similar to the Förster law corresponding to disorder of the optically-active ions in the medium, yet differs in the value of the coefficient in the expression for the macroparameter γ_H (back transfer taken into account).

It is clear from Expression (55) that the length of the ordered, exponential decay region ($t < t_1$) is substantially greater for media with weak interactions, a small value of C_{DD} and a significant minimum possible distance R_{min}. Here the luminescence decay level in the ordered section, as demonstrated in Section 1, is determined by the active ion concentration and decays sharply as n approaches the maximum possible activator concentration given by $n_{max} = \sqrt{2}R_{min}^{-3}$.

A number of theoretical studies have assumed a diffusion-like character to energy migration in the last (third) stage for $t > t_2$; the precise form of the kinetics of this stage had not been established until recently.

The great value in investigating energy migration in a spectrally and spatially inhomogeneous ensemble of particles from the general physical viewpoint and the extreme scarcity of experimental results require developing mathematical and direct experimental techniques for investigating their kinetic mechanisms, stages, and transition ranges, and for establishing the relations between the macro- and microcharacteristics of the process.

Computer statistical modeling of excitation migration throughout a disordered particle ensemble. Given the difficulties of obtaining an exact analytic solution of a system of coupled rate equations describing energy migration

$$dP_i(t)/dt = \sum_j [W_{ji} P_j(t) - W_{ij} P_i(t)] \tag{164}$$

for the entire time interval a number of studies were carried out [112, 156] where the excitation migration process was investigated by computer statistical modeling. These studies demonstrated a good agreement between the approximate analytical solution $P(t) = \exp(-\sqrt{W_N t})$ for the disordered stage with the "exact" solution obtained by modeling over rather short periods.

However, Refs. [112, 156] analyzed only the disordered stage and neglected the latter stage of migration kinetics. Representing the data from Refs [112, 156] in graph form complicates their use for analyzing all kinetic stages. Moreover, the studies were carried out for dipole-dipole donor interaction only (low concentrations $c = 0.01, 0.2$) and for interaction between nearest neighbors at high concentrations ($c = 0.5, 0.75, 1.0$). The maximum deactivation level of the initial center in these studies $I(t) = P(t) > 0.01$ and

$P(t) > 0.02$ is not always sufficient for a thorough analysis of the diffusion stage.[10]

We therefore carried out statistical modeling of excitation migration using the method developed in Refs [95, 112] for dipole-quadrupole and quadrupole-quadrupole donor-donor interactions.

Excitation migration was investigated in an ensemble of impurity centers randomly distributed among the sites of the crystalline lattice. The quantity under analysis was the probability of finding excitation at the initially-excited center (the donor). The donor-donor excitation transfer rate depended solely on the distance between donors: $W_{ij} = C_{DD}/R_{ij}^S$, where S is the multipole order of donor-donor interaction. This study therefore has modeled selective laser excitation and spectral-nonselective excitation migration along an inhomogeneously-broadened line generated by reversible multipole-multipole interaction between donors.

The present study investigated excitation migration kinetics (the decay kinetics of the initial site) in the specific concentration range $c = 0.01-0.75$.

The excitation migration modeling technique used here was proposed in Ref [95] and was based on obtaining a solution of a system of coupled rate equations that has already been averaged over all possible initially-excited donors for the selected donor configuration:

$$P(t) = \frac{1}{N} \sum_{i=1}^{N} P_i^{(i)}(t),$$

$$(165)$$

where N is donor number; the superscript in parentheses represents the number of the initially-excited donor, i.e., $P^{(i)}(t)$ is the solution of System (164) with the initial conditions $P_k^{(i)}(0) = \delta_{ik}$. We write (164) as

$$dP_k^{(i)}/dt = \sum_j \Gamma_{kj} P_j^{(i)}(t),$$

$$(166)$$

where $\Gamma_{ii} = -\Sigma_j W_{ij}$ and $\Gamma_{ij} = W_{ij}$ for $i \neq j$.

Following Ref. [95] we can obtain

$$P(t) = \frac{1}{N} \sum_\alpha \exp(-\gamma_\alpha t),$$

where δ_α are the eigenvalues of matrix Γ.

Therefore the process of obtaining the solution of System (166) averaged over all possible initially-excited centers reduces to finding the

---

[10] The quantity $P(t)$ investigated in the present section for the population of the initial site is adequate for our values of $I(t)$ and $I(E_L, t)$ introduced previously.

eigennumbers of the system matrix Γ. An advantage of this method of solving this system of rate equations compared to direct numerical integration of System (166) is that it results in a solution that is already averaged over all possible initially-excited donors which reduces the statistical spread of the model kinetics.

The limited computer RAM capacity makes it necessary to limit the number of donors N and the dimensions of the model crystal l (henceforth we shall also refer to it as a cube) since $N(N + 1)/2$ RAM words are required to store matrix Γ in the RAM accounting for the symmetry of the matrix. Therefore the limit on the number of donors in this study was $N \leqslant 800$.

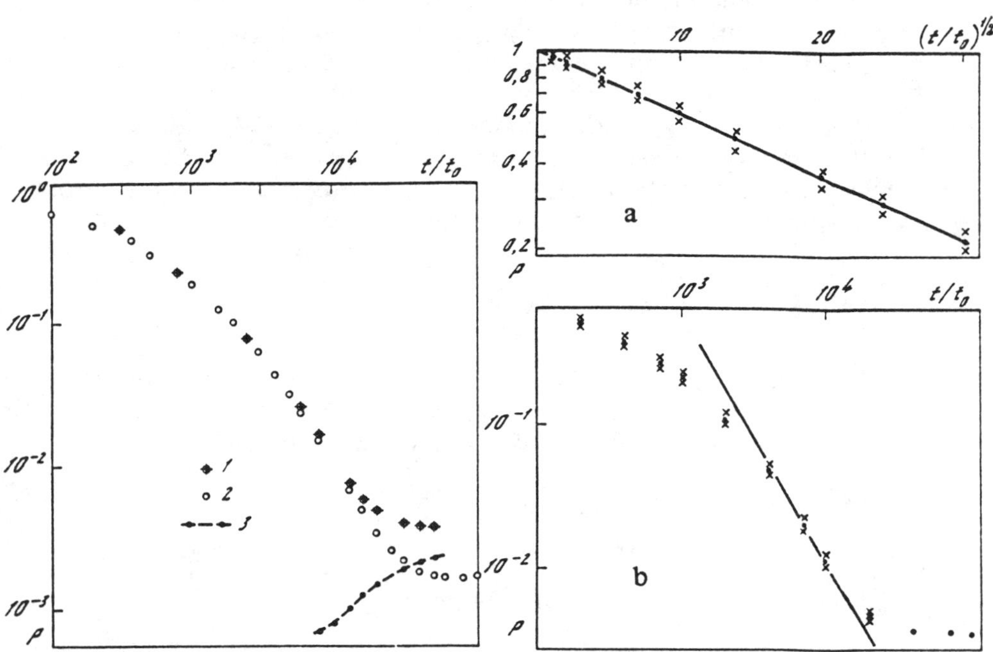

Fig. 23. Analysis of the effect of a finite number of donors.
$1 - P_{270}(t)$; $2 - P_{640}(t)$; $3 - P_{270}(t) - P_{640}(t)$.

Fig. 24. The effect of the statistical spread of the kinetics for various configurations.
a – disordered stage; b – migration over long time periods.

The finite cube size l and number of donors N affect the model kinetics in two ways. First excitation will, over time, be "mixed" uniformly between the donors in a system with a finite number of donors and therefore the probability of finding excitation at the initially-excited donor (population of the initial site) approaches $1/N$ as $t > \infty$ in the absence of excitation decay. Mathematically this means that 0 is the eigennumber of the matrix Γ. In actual systems the number of donors is extremely high ($>10^{18}$) and in practice $P(t)$ approaches 0 as $t \to \infty$ at the same time that in the present study due to the few donors N the convergence of $P(t)$ on $1/N$ is responsible for the primary error in the modeling result over long periods and determines the ultimate range of the population of the initial site $P(t)$.

In order to estimate the effect of donor number the calculation was carried out for two different donor configurations having an identical concentration $c = 0.01$ yet a different donor number: $N = 270$ and $N = 640$. The calculation was run for dipole-dipole interaction between donors as the most long-range of the types of interaction considered.

The calculation relations $P_{270}(t)$ and $P_{640}(t)$ as well as the difference $P_{270}(t) - P_{640}(t)$ were plotted on a common graph on coordinates $\lg t$, $\lg P$ for the analysis (Figure 23). It is clear that the kinetics of $P_N(t)$ asymptotically approach $1/N$ as $t \to \infty$. The error of $P_{270}(t) - P_{640}(t)$ diminishes noticeably over time and is less than 10% of $P_N(t)$ as early as times when $P_N(t) = 3/N$ rather than $10/N$ which would be the case if the error level were constant. It also follows from the time dependence of the error level that this error cannot be eliminated by simple subtraction of $1/N$ from $P_N(t)$. Evidently the increase in error with time is related to the fact that excitation is more strongly localized over short time periods and excitation migration is subject to the lesser effect of the cube edges.

We can conclude from this analysis that the error, i.e., the difference $P_N(t) - P_\infty(t)$ between the solutions for the population of the initial site for a system of N donors and a system with a very large number of donors is rather small until $P_N(t) > 3/N$.

Another aspect of the effect of initial cube dimensions is the difference in the excitation transfer conditions in the center and along the cube edges. Periodic boundary conditions were used to eliminate the effect of the cube boundaries, specifically: if the difference of the coordinates of any two donors $\Delta x_{ij} > l/2$, where l is the length of the cube side, this difference is substituted with $\Delta x_{ij} = l - \Delta x_{ij}$. It is this quantity $\Delta x'_{ij}$ that is used to calculate the difference between donors: $r_{ij} = (\Delta x'^2_{ij} + \Delta y'^2_{ij} + \Delta z'^2_{ij})^{1/2}$. Physically this means that the model cube is a "superlattice" cell of period l where the donor configuration and initial excitation conditions are translated with a period l along the coordinate axes and interaction between the excited donor and only those donors within the "superlattice" cell containing the excited donor at the center will be accounted for.

A calculation was carried out for 52 different donor configurations for an identical concentration $c = 0.01$ and an identical donor number $N = 270$ for dipole-dipole interaction between donors in order to analyze the effect of the statistical spread of migration kinetics for different donor configurations. These calculation results were then averaged. The average, maximum, and minimum populations of the initial site were then plotted (Figure 24).

It is clear from Figure 24 that all kinetics manifest approximately identical functional behavior where the deviation from the mean (circles) does not exceed 10-12% which yields satisfactory modeling accuracy even without averaging over the configurations. The greatest spread of functional behavior of the kinetics is observed in the initial ordered migration stage which is evidently due to the stronger effect of local inhomogeneities of the spatial distribution of the donors nearest the initial site in this stage. The values of the disordered stage macroparameters W_H for the various configurations differ only slightly and the majority of the spread in this stage can be attributed to the spread of deactivation levels of the initial site in the initial ordered stage. The excitation migration rate in the diffusion stage also varies moderately for the different configurations.

A computer program was written to model excitation migration; this program carried out the following successive operations.

1. Introduction of parameters: specific concentration c, interaction multipolarity S, cube dimensions l.

A simple cubic lattice was selected as the crystalline lattice of the model crystal for simplicity. The lattice R_{min} was selected as the unit of length. The inverse of the excitation transfer rate over the distance of the lattice period was selected as the unit of time: $t_0 = R_{min}^S / C_{DD}$. The donor concentration n and specific concentration c are numerically identical in these units.

2. Calculation of donor number $N = cl^3$.

3. Configuration generation. A pseudorandom number generator is used to determine the numbers of the crystalline lattice sites containing the donors. In order to obtain different donor configurations a specific (given in the input data) number of resets are incorporated in the random generator prior to use.

4. Calculation of the elements of system matrix (166) Γ_{ij}.

5. Reduction of matrix Γ to diagonal form by means of Jacobi iteration rotations. This method was selected in connection with the structure of matrix Γ which contains multiple chaotically-distributed, very small elements corresponding to the transfer rates between distantly-separated donors. These elements can be neglected in calculation of the eigennumbers which in this method occurs automatically.

6. Print-out of results.

As noted above the kinetics of excitation migration in the donor system can be divided into three stages described by the following approximation formulae:

$$\text{ordered} - P(t) = \exp\left(-W_{\max}t\right), \quad t < t_1, \tag{167}$$

$$\text{disordered} - P(t) \sim \exp\left[-(W_{\mathrm{H}}t)^{3/S}\right], \quad t_1 < t < t_2, \tag{168}$$

$$\text{diffusion} - P(t) = (W_{\text{диф}}t)^{-3/2}, \quad t > t_2. \tag{179}$$

It is interesting to estimate the time to the diffusion stage in the decay kinetics. For the case of moderately high donor concentrations ($c \leqslant 0.2$) when diffusion results in replacement of nonstationary kinetics of the (51) type the boundary time t_2 can be found from equality of the derivatives $d[\lg I(t)]/d(\lg t)$ for substitutional decay laws (51) and (35) [125]:

$$W_{\mathrm{H}}^{3/S}(3/S)(t_2)^{3/S} = 3/2, \tag{170}$$

from here

$$t_2 = (S/2)^{S/3} W_{\mathrm{H}}^{-1}. \tag{171}$$

It is clear from the last expression that

$$\text{for } S = 6 \quad t_2 = 9W_{\mathrm{H}}^{-1},$$
$$\text{for } S = 8 \quad t_2 = 40W_{\mathrm{H}}^{-1}, \tag{171a}$$
$$\text{for } S = 10 \quad t_2 = 213W_{\mathrm{H}}^{-1},$$

This suggests a significantly later onset of the diffusion stage with increasing multipole order of interaction even at a fixed average transfer rate, which correlates with the result of Ref. [117].

We note that the estimate for $t_2 = 9W_{\mathrm{H}}^{-1}$ is similar to the estimate we obtained previously based on the results of Ref. [114] (see Expression (63)): $t_2 = 14.5W_{\mathrm{H}}^{-1}$ and substantially exceeds $t_2 = 0.5W_{\mathrm{H}}^{-1}$ obtained in Ref. [117] (see Expression (72)).

In analyzing the modeling results we selected the three excitation migration kinetic stages identified above and investigated the dependence of the macroparameters $W_{\max}$, W_{H}, and W_{diff} in these stages on donor concentration c and donor-donor interaction multipolarity S as well as the dependence on c and S of the boundary times t_1 and t_2 of the transition from one stage to another.

The technique examined in Ref. [25] was used to analyze the results. The computer-calculated kinetics $P(t)$ were plotted on rectified coordinates in order to achieve a linear relation on these coordinates and therefore easily determine the excitation migration macroparameters at the stages under investigation.

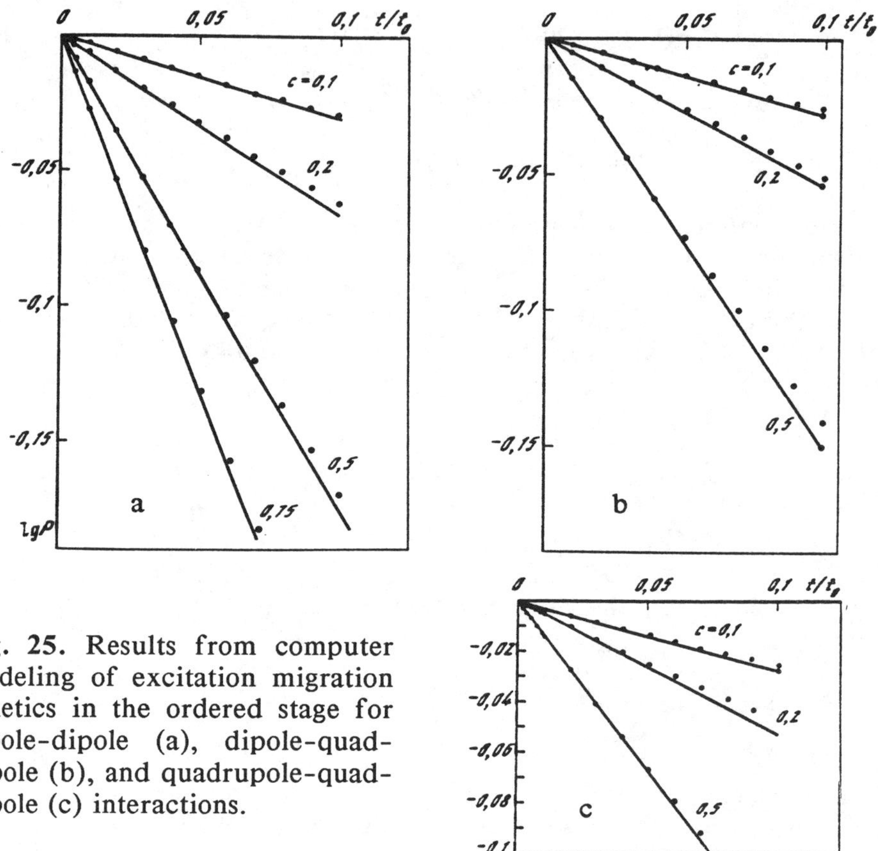

Fig. 25. Results from computer modeling of excitation migration kinetics in the ordered stage for dipole-dipole (a), dipole-quad-rupole (b), and quadrupole-quad-rupole (c) interactions.

For the ordered stage these coordinates are t and $\lg P$ (the approximation formula of the first stage $\lg P = -\lg\ eW_{max}t$), and for the disordered stage $t^{3/S}$, $\lg P$($\lg P = -\lg\ eW_H^{3/S}t^{3/S}$), while for the diffusion stage $-t^{-3/2}$, P($P = W_{diff}^{-3/2}t^{-3/2}$). In order to find the boundary times t_1 and t_2 a graph of the $P(t)$ kinetics was plotted on $\lg t$, $\lg P$ coordinates to analyze the boundary time to the diffusion stage t_2 and on $\lg t$, $\lg(-\lg(P))$ coordinates to analyze the boundary time of the ordered stage t_1.

The initial, ordered (or exponential) stage becomes significant only in crystals with a high specific donor concentration: $c \geqslant 0.1$. A rectilinear section is clearly evident in the graphs of $P(t)$ on coordinates t, $\lg(P)$ (Figure 25) indicating the ordered stage is present for all cases under analysis.

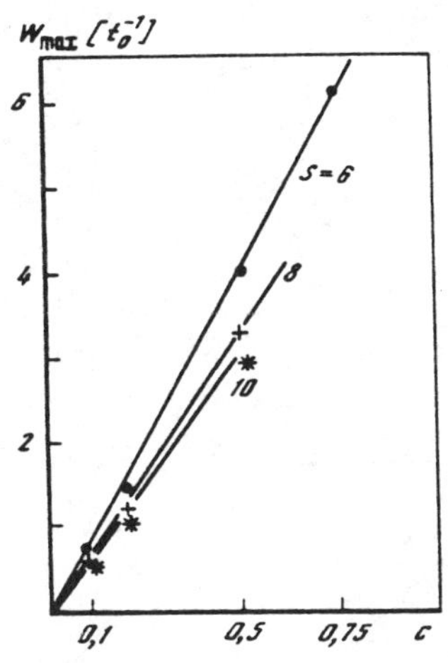

Fig. 26. Initial stage migration rate plotted as a function of concentration.

Analysis revealed that the transfer rate in the ordered stage W_{max} is linearly dependent on concentration c in the dipole-dipole, dipole-quadrupole, and quadrupole-quadrupole interaction cases (Figure 26). This is consistent with Formula (49) $W_{max} = cC_{DD}\Sigma_i R_{0i}^{-S}$, where summation is carried out over all lattice sites. The rate values W_{max} (in units of t_0^{-1}) are given in Table 2.

Table 2

Variation in the maximum migration rate W_{max} as a function of c for S = 6, 8, and 10.

S	c			
	0.1	0.2	0.5	0.75
6	0.76	1.54	3.91	6.25
8	0.62	1.26	3.38	–

S	c			
	0.1	0.2	0.5	0.75
10	0.62	1.24	3.11	–

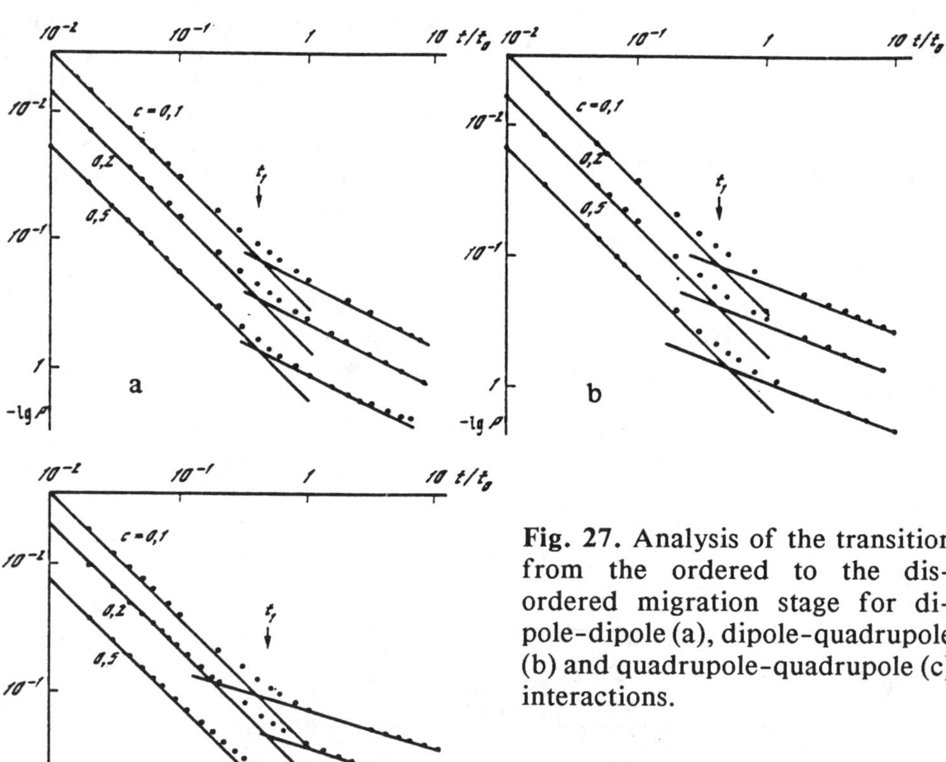

Fig. 27. Analysis of the transition from the ordered to the disordered migration stage for dipole-dipole (a), dipole-quadrupole (b) and quadrupole-quadrupole (c) interactions.

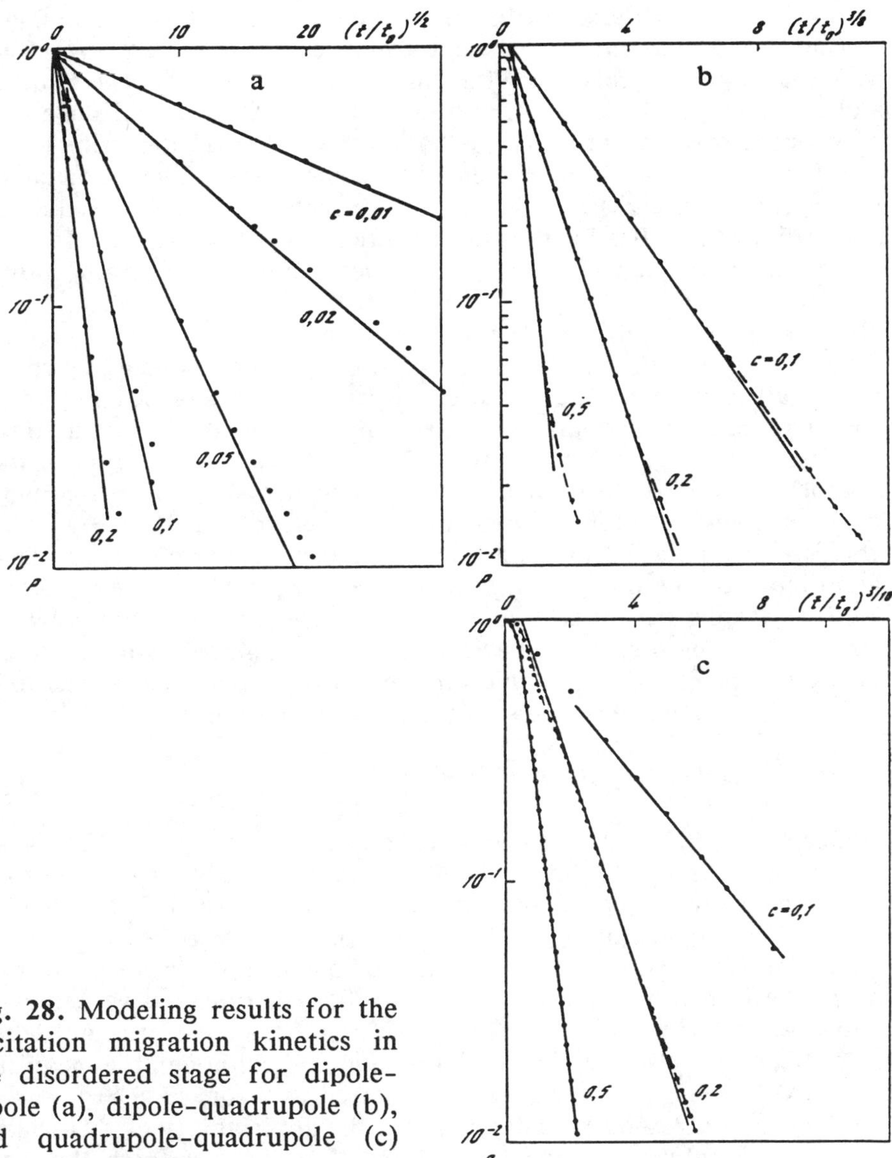

Fig. 28. Modeling results for the excitation migration kinetics in the disordered stage for dipole-dipole (a), dipole-quadrupole (b), and quadrupole-quadrupole (c) interactions.

The computer-calculated kinetics were plotted on coordinates $\lg t$, $\lg(-\lg P)$ to determine the boundary time of the ordered stage (Figure 27). On these coordinates the kinetics are accurately approximated by a dashed line whose slopes are equal to one for the ordered stage and $S/3$ for the disordered stage of excitation migration kinetics. Using the time corresponding to the apex of the broken line curve as the boundary time t_1 of the transition from the ordered to disordered stage we see that $t_1/t_0 \simeq 0.4\text{-}0.5$ for all moderately large donor concentrations: $S = 6$ and $c \leqslant 0.5$ for $S = 8$, 10 which is in good agreement with Formula (55): $t_1 = t_0/2$. The boundary time clearly corresponds to a direct transition from the ordered stage to the diffusion stage for concentration $c \geqslant 0.5$ in dipole-dipole interaction ($S = 6$).

The $P(t)$ kinetics were calculated by computer and plotted on coordinates $t^{3/S}$, $\lg P$ (Figure 28) in order to analyze the second, disordered excitation migration stage. The kinetics of excitation migration were analyzed for donor systems with the following donor concentration values c and interaction multipolarities: $c = 0.01, 0.02, 0.05, 0.1, 0.2$ for $S = 6$; $c = 0.1, 0.2, 0.5$ for $S = 8$; and $c = 0.1, 0.2, 0.5$ for $S = 10$.

The graph of the $P(t)$ kinetics on these coordinates undergoes rectification for all parameters indicated, suggesting a disordered excitation migration stage in these systems. With large times a deviation from a straight line is observed, suggesting the onset of the next excitation migration stage. This "root" stage, if it exists, will occur only over a very short time period which is clear in the graph on the coordinates $\lg t$, $\lg(-\lg P)$ (see Figure 27a) for the case of excitation migration in a system of donor concentration $c = 0.5$ and dipole-dipole interaction between the donors.

The graphs were used to determine the average excitation migration rate in the disordered stage W_{H}. The values of W_{H} coincide with those predicted by theory for dipole-dipole interaction for all concentrations in the range $c = 0.01\text{-}0.2$ and for dipole-quadrupole interaction for $c = 0.1$:

$$W_{\mathrm{H}} = [(4\pi/3)\Gamma(1 - 3/S)\, n \cdot 2^{1-3/S}]^{S/3} C_{\mathrm{DD}}. \tag{172}$$

The value $\gamma_{\mathrm{H}} = W_{\mathrm{H}}^{3/S}$ for dipole-quadrupole interaction for $c = 0.2$ is 14% greater, while it is 18, 30, and 37% greater than that obtained by Formula (172) for quadrupole-quadrupole interaction for $c = 0.1, 0.2,$ and 0.5, respectively. The growth of W_{H} and γ_{H} for dipole-quadrupole interaction in the transition from $c = 0.1$ to $c = 0.2$ and for quadrupole-quadrupole interaction in the transition from $c = 0.1$ to $c = 0.5$ is in qualitative agreement with theoretical predictions describing the nonlinear concentration dependence of W_{H} and γ_{H} (see Section 1, Expression (57)) although the magnitude of this growth is substantially greater compared to theory (3-9%).

The data obtained here demonstrate the applicability of this analysis and of simple analytic expressions (51) and (52) for describing the disordered energy migration stage not only for $c \ll 1$ but also for relatively high donor concentrations up through $c \approx 0.2$.

The structure of the known diffusion solution for energy migration over an ordered lattice $I(t) = (W_{\text{diff}}(t))^{-3/2}$ predicts that it is best to search the diffusion decay stage by plotting the kinetics on coordinates $\lg P$ as a function of $\lg t$. Here the presence of the diffusion-like stage can be established based on observation of the linear decay kinetic stage (plotted on these coordinates) with a slope of $-3/2$.

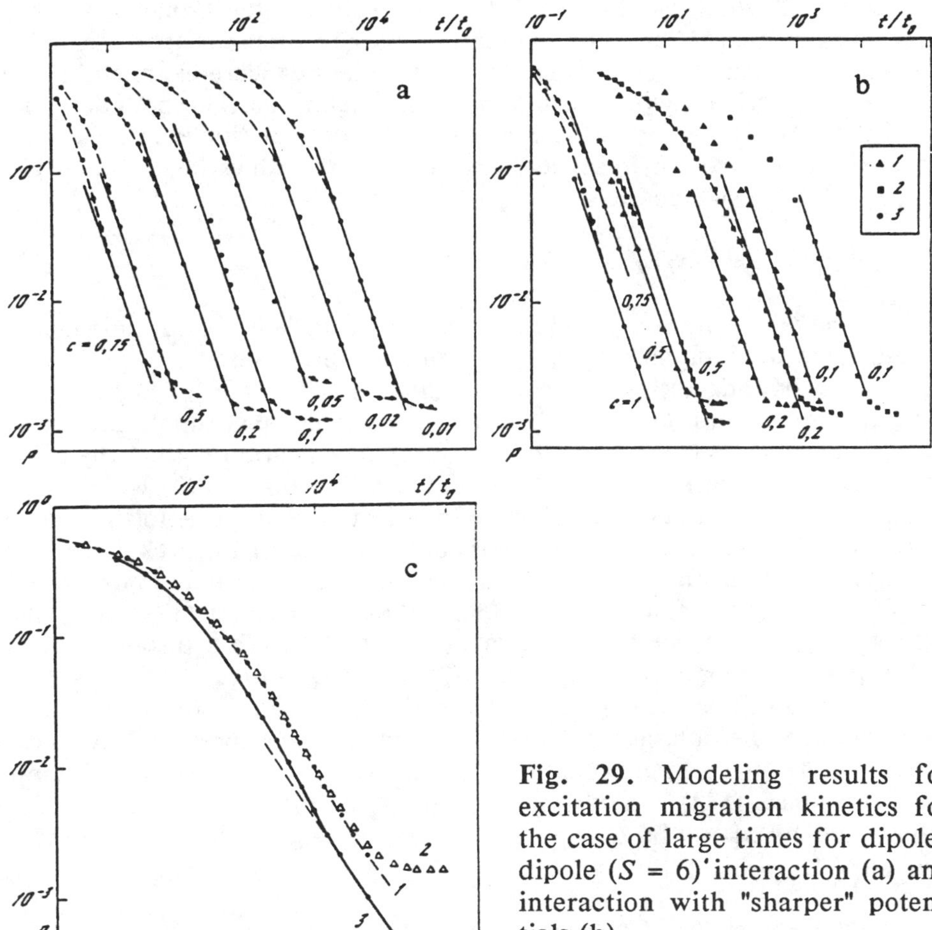

Fig. 29. Modeling results for excitation migration kinetics for the case of large times for dipole-dipole ($S = 6$) interaction (a) and interaction with "sharper" potentials (b).

b: 1 - $S = 8$, 2 - $S = 10$, 3 = short-range potential; c: calculation by approximation formula.

The computer-calculated $P(t)$ relation for each of the donor configurations was plotted on coordinates $\lg t$, $\lg P$ (Figure 29a, b) in order to analyze the excitation migration kinetics over long time periods. All calculated kinetics with sufficiently long time periods are accurately approximated by straight lines with slopes of $-3/2$ which can be treated as the onset of the diffusion migration stage in a disordered donor set where $P \sim t^{-3/2}$.

An overvibration is detected in the dipole-dipole interaction case for rather high concentrations ($c > 0.2$) in the prediffusion stage of the $P(t)$ graph for $P \sim 0.03-0.05$ on these coordinates. The overvibration turned out to be stronger the higher the donor concentration c although the slope does not exceed -2 which would be the case according to the Dzheparov theory [122] (see Formula (102)). The population of the initial site tends to $1/N$ as $t \to \infty$; the reasons for such behavior of the kinetics were examined previously in the section devoted to the modeling technique. The boundary time for the crossover to the diffusion stage for dipole-dipole interaction at concentrations $c < 0.2$ is in satisfactory agreement with estimate (171) [125]: $t_2 = (S/2)^{S/3} W_H^{-1}$, which yields

$$P(t_2) \approx \exp(-S/2). \tag{173}$$

The overvibration noted above makes it difficult to use this method to determine the boundary time t_2 at high concentrations.

The boundary time for concentrations $c = 0.1$ and $c = 0.2$ is also in good agreement with (173): $P(t_2) \sim 0.02$ for dipole-quadrupole interaction between the donors (see Figure 29b). For a concentration $c = 0.5$ the diffusion stage occurs earlier than that predicted by Formula (173) which is due to the diminishing deactivation level of the initial center due to the significant ordered stage. It is interesting that the migration kinetics can be described by a diffusion law for quadrupole-quadrupole interaction even with the boundary population $P(t_2) \sim 0.05$. Although Formula (173) suggests a significantly later onset of diffusion $P(t_2) \sim 0.007$ such a discrepancy lies within the modeling accuracy and is not subjected to precise analysis.

A good description of the migration kinetics in dipole-dipole donor interaction can be obtained simultaneously in the disordered and the diffusion stages from an approximation formula which is obtained if Formula (102) from Ref. [122] is modified to account for the back migration in the Hüber model [95] and by dropping the term making a contribution proportional to t^{-2}:

$$P(t) = \exp[-W_H t)^{1/2}] + \{1 - \exp[-W_H t)^{1/2}]\}[W_{\text{diff}}(t + \tau)]^{-3/2},$$

where τ is the parameter selected for joining of the two kinetic stages.

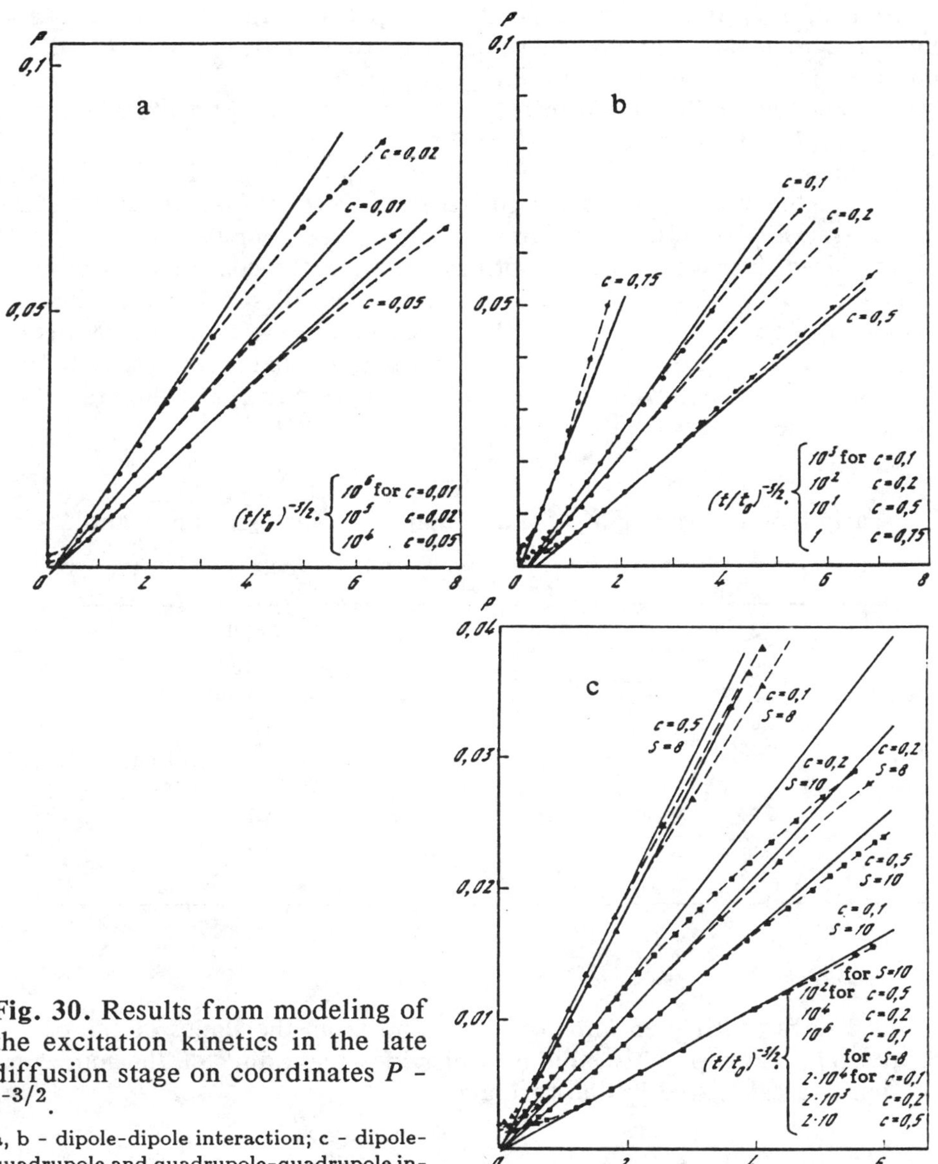

Fig. 30. Results from modeling of the excitation kinetics in the late diffusion stage on coordinates $P - t^{-3/2}$.

a, b - dipole-dipole interaction; c - dipole-quadrupole and quadrupole-quadrupole interactions.

The calculation result using this formula is given in Figure 29c (curve 1). It is clear that the modified expression (dashed curve) accurately describes the numerical modeling result (curve 2) over the entire range of variation of $P(t)$ at the same time that Dzepatov Formula (102) (curve 3) yields a significant error that increases with time. This can be treated as a poor accounting of back excitation to the initially-excited donor in the Dzeparov model [122] and suggests that the experimental results confirming this in Ref. [157] are in error.

In order to determine the average migration rate in the diffusion stage W_{diff} the computer-calculated kinetics were plotted on coordinates $t^{-3/2}$, P (see Figure 30). The migration kinetics $P(t)$ for $t \to \infty$ ($t^{-3/2} \to 0$) are linearized, although they do not pass through the coordinate origin but rather intersect the Y-axis at the point with the smallest relative ordinate: δ. The quantity $\delta \approx 1/N$ (N is the number of donors in the cube) also is weakly dependent on concentration which makes it possible to attribute such behavior of the model kinetics to the effect of the finite number of donors, which is unavoidable in statistical modeling.

Table 3

Variations in the rate W_{diff} and parameter α for different c and S

c	W_{diff}	α	c	W_{diff}	α
	$S = 6$			$S = 10$	
0.1	$1.9 \cdot 10^{-3}$	1.5	0.1	$5 \cdot 10^{-3}$	0.85
0.02	$7.6 \cdot 10^{-3}$	1.52	0.2	$6.1 \cdot 10^{-2}$	1.03
0.05	0.046	1.48	0.5	1.71	1.35
0.1	0.19	1.5		Short Range Potential	
	$S = 8$				
0.1	0.03	1.13	0.75	5.7	–
0.2	0.205	1.19	1.0	12.56	1.0
0.5	2.78	1.4			

The rates W_{diff} (in units of t_0^{-1}) found from the slopes of the curves are given in Table 3. This table also provides the values of the numerical coefficient α calculated by the formula

$$W_{diff} = 4\pi n^{2/3} D = 4\alpha\pi C_{DD} n^{S/3} = 4\pi\alpha c^{S/3} t_0^{-1}.$$

A sharp drop off in the migration rate (by a factor of ~40, quasi-localization) with increasing slope of the interaction potential in the disordered system is clearly evident from the table (dilute crystal, $c = 0.1$) which is not observed at high concentrations. The constant value of the coefficient $\alpha = 1.55 \pm 0.07$ for different c in dipole-dipole interaction ($S = 6$) confirms the square-law dependence of the migration rate on concentration, indicating an irrational degree of dependence of the diffusion coefficient on concentration $D = \alpha C_{DD} n^{4/3}$.

The value of the numerical parameter α which is important to theory and has been found by statistical modeling ($\alpha = 1.55$ for $S = 6$) was below the values of α obtained in a variety of theoretical studies: 3.38 in Refs. [23, 111] (estimate, 2.2 in Ref. [114]), and $\alpha = 3.42$ in Refs. [116, 117] neglecting back transfer and 2.72 taking into account back transfer while $\alpha = 3.26$ in Ref. [115] accounting for back transfer in a two particle approximation and 2.88 in a three particle approximation, $\alpha = 2.65$ in Ref. [118] and 3.46 in Ref. [122]. The quantity α diminishes, as we see from Table 3, for dipole-quadrupole and quadrupole-quadrupole interactions, corresponding to theoretical expressions (74) and (75) and attributable to the minor contribution of interactions with long-range coordination shells. However, the numerical values of α (for $S = 8$ and 10) at low concentrations ($c \leqslant 0.2$) obtained by the Monte-Carlo method were below the theoretical values of α as in the case $S = 6$ for low concentrations [118]: 2, 3 for $S = 8$, and 1.14 for $S = 10$.

The fact that accounting for back transfer reduces the diffusion coefficient and the value of α permits us to treat the elevated values of α_{th} over those we have found as incomplete incorporation of back transfer in existing theoretical models.

We will expand this picture with an analysis of the migration kinetics for the case of a short-range interaction potential (SRP) when ion-ion interaction responsible for excitation transfer is activated solely between nearest neighbors in the lattice at the minimum possible distance R_{min}. Analysis of an exact solution for the case of migration throughout an ordered cubic lattice ($c = 1$, see Expression (39)) and kinetics for $c = 0.75$ from Refs. [95, 112][11] reveals that their graphs on coordinates $\lg P(\lg t)$ (see Figure 29) and $P(t^{-3/2})$ (see Figure 30) are linearized with sufficiently long time periods. This suggests a diffusion character to migration without excitation localization for the short-range (contact) interaction potential as well at concentrations exceeding the percolation concentration $c > c_{per} = 0.307$. An overvibration preceding the diffusion stage is clearly visible in Figure 29 in deriving an exact solution for the short-range potential case ($c = 1$), although the nature of this overvibration is different than that assumed by

---

[11] Migration kinetics for $c = 1.0$ and $c = 0.75$ under short-range conditions were presented in Refs. [95, 112]. We carried out the first analysis of the kinetics in which the diffusion stage of migration was detected for $c = 0.75$ in Ref. [125].

the authors of Ref. [122] where it appeared only for the longest range interaction: $S = 6$.

A comparison of the migration kinetics with changing slope of the interaction potential ($S = 6$, 8, and 10 and the short-range potential) and degrees of disordered (concentration c) shows that the overvibration amplitude is maximized for $S = 6$ (see Figures 29, 30, $c = 0.75$, 0.5). A reduction in concentration (a rise in disorder) attenuates the overvibration amplitude and in the final analysis will result in complete decay (see Figures 29, 30, $c \leqslant 0.2$). A rise in the potential slope will also attenuate the overvibration (see Figures 20, 30, $S = 8$, $c = 0.5$). The joint effect of the short-range potential and disorder will cause the strongly-expressed overvibration at $c = 1.0$ under short-range conditions to escape detection even with $c = 0.75$ (see Figures 29, 30) while the diffusion stage begins at $t_2 = 0.74t_1$ and continues through a variation in intensity exceeding an order of magnitude.

Experimental investigations of the ordered and disordered stages of the migration kinetics for quadrupole-quadrupole interaction of Eu^{3+} *[125].* The first and second energy migration stages identified above are most easily investigated experimentally by observing spectral energy migration among Eu^{3+} ions in glass. We have demonstrated in Refs. [84, 108] that high-temperature ($T = 300$ K) spectral migration among Eu^{3+} ions in Na-Y-B-silicate glass at a concentration $n = 10^{21}$ cm^{-3} is characterized by a quadrupole-quadrupole interaction mechanism ($S = 10$) at an average rate several times smaller than the radiative deactivation rate of the metastable level. The absence of substantial dispersion of the radiative probabilities and luminescence decay rates for the different optical centers substantially simplifies the analysis of the spectral migration dynamics for this object.

Segments of the luminescence spectra of Eu^{3+} ions at the $^5D_0-^7F_0$ transition are shown in Figure 31a after selective resonant excitation at this same transition by means of a narrowband ($\Delta v < 1$ Å) tunable dye laser (Rhodamine 6Z) pumped by the second harmonic of a YAG:Nd^{3+} laser (pulse duration $t_p = 10^{-8}$ s, repetition rate $f = 12.5$-50 Hz). It is clear that a strong energy redistribution between the narrow ("donor") and broad ("pedestal") bands towards the broad band is observed with increasing delay.

Figure 31b shows the luminescence decay kinetics of Eu^{3+} ions recorded by the double selection technique when the selective recording wavelength was set equal to the selective excitation wavelength: $\lambda_R = \lambda_E = 576$ nm. The kinetics were measured by means of a DFS-12 spectrometer with a 5 Å resolution by single-channel recording and averaging (time resolution: 1.0 μs) on a PAR-162 strobe-integrator and by multichannel recording in real time (1024 channels with a time resolution of 5 μs per channel) followed by summing of 100-500 decay kinetics on the PAR-4202 averager. It is clear from a comparison of the decay kinetics and the varying integral intensity of the narrow donor peak in the luminescence spectra that it provides a good reflection of the decay kinetics of the "donor" centers only in the initial stage of the migration process. At a later stage of spectral migration luminescence from the broad band of the "pedestal" makes an increasingly large contribution to luminescence at $\lambda_R = \lambda_E$, which complicates the analysis process.

We also believe calculating the decay kinetics of the donor centers based solely on time-resolved spectra is not completely beyond reproach since it is wholly based on the selection of the spectral normalization method as well as the accuracy of the spectral analysis of the complex luminescence contour.

We therefore developed an analysis technique based on a comparison of kinetic and spectral data. The spectra measured at different times are normalized here based on the resonance luminescence decay kinetics measured by the double selection method. A contour corresponding to the transmission spectrum of the DFS-12 spectrometer in the experiment to measure decay kinetics by double selection was added to the luminescence spectra for this purpose, as demonstrated in Figure 31a for $t_D = 3$ ms. The area of this contour S_s corresponds to the intensity $\tilde{I}(t)$ in the luminescence decay kinetics and is normalized to determine the integral intensity $I^*(t)$ of the narrow "donor" peak of area[12] S_R.

The donor luminescence decay kinetics measured by this technique has a better degree of reliability.

As demonstrated above a convenient technique for isolating and analyzing the various stages of energy transfer kinetics is to plot $I(t) = I^*(t)/I_{rad}(t)$ on coordinates $\lg[-\lg I(t)]$ as a function of $\lg t$. The slope of these relations yields the order of the parameter t, thereby permitting differentiation of the ordered (exponential) decay stage $(\sim t)$ from the disordered (nonexponential) stage $(\sim t^{3/S})$ and to determine the multipolarity of the interactions for the latter on the exponent $3/S$.

It is clear from Figure 32a that kinetics represented on coordinates $\lg[-\lg I(t)]$ - $\lg t$ are in fact described by a straight line in the beginning of the process for $t < t_1 = 60$ μs; the slope of this line $\mathrm{tg}\varphi = 1$, while in the final stage of the process (for $t > t_1$)$\mathrm{tg}\varphi \simeq 3/10$ $(S = 10)$ corresponding to nonstationary energy migration through a disordered set of ions in quadrupole-quadrupole interaction. It is also important to identify the transition range near t_1 where the slope varies continuously from 0.3 to 1, running continuously over all intermediate values.

---

[12] Measuring $I^*(t)$ based on the area rather than the intensity of the narrow spectral component makes it necessary to compensate the effect of the spectrally-selective resonant migration process $Eu^{3+} \rightleftharpoons Eu^{3+}$ on the desired kinetics; this serves to wash out (broaden) the narrow "donor" component.

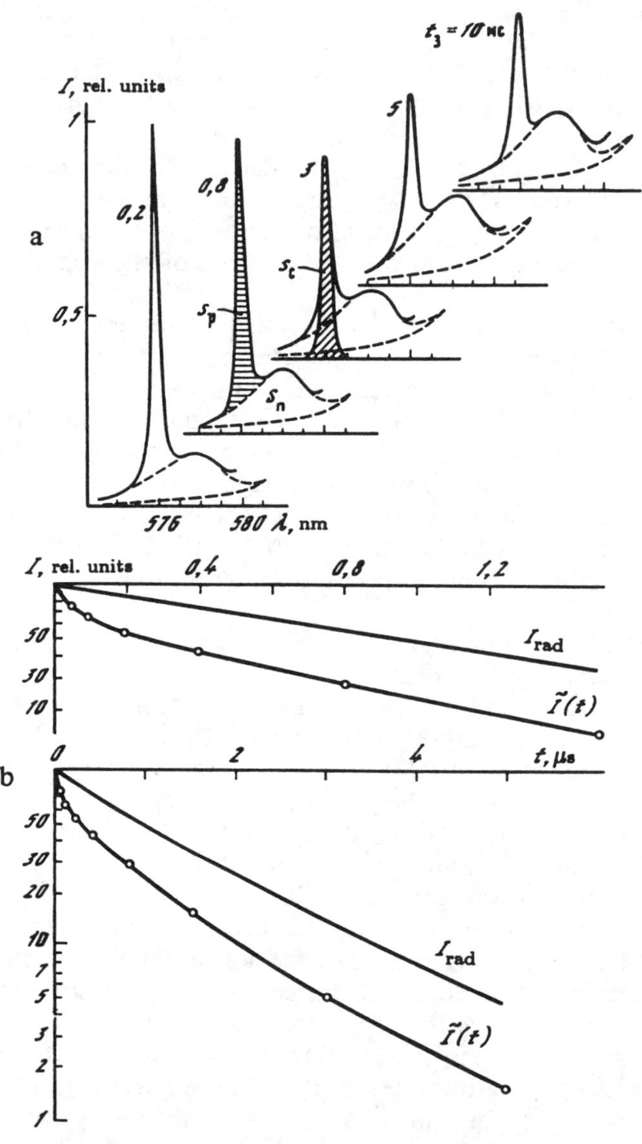

Fig. 31. Spectral-time dynamics of energy migration within the inhomogeneously-broadened $^5D_0\!-\!^7F_0$ luminescence band of Eu^{3+} in Na-Y-B-silicate glass (c = 12%, T = 300 K) (a) and the kinetics of spontaneous $I_{em}(t)$ (c = 2%, T = 77 K) and migration-accelerated $\tilde{I}(t)$ (c = 12%, T = 300 K) luminescence decay measured by the double selection method (b).

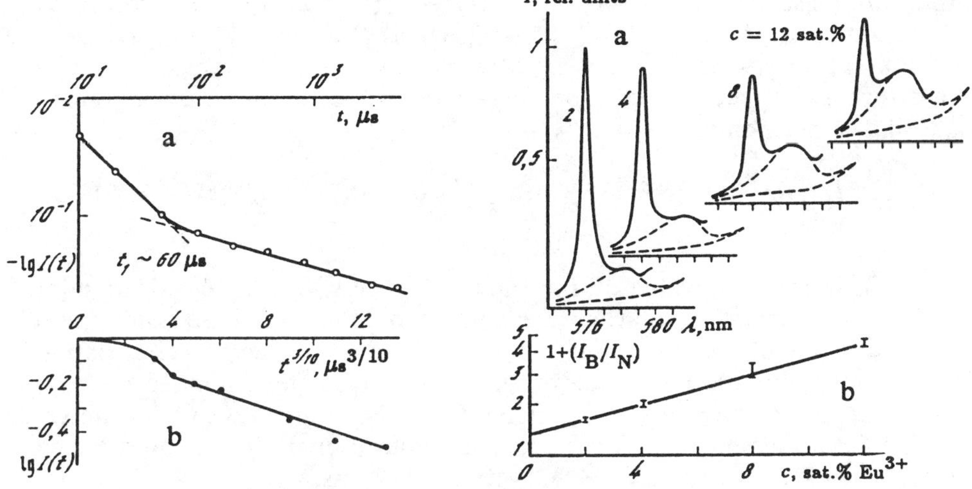

Fig. 32. Kinetics for the ordered ($t < 60$ μs) and disordered ($t > 60$ μs) energy migration stages on coordinates lg($-$lg I) $-$ lg$t$ (a) and lg$I$ $-$ $t^{3/10}$ (b).

Fig. 33. Concentration variations in the luminescence spectra of Eu^{3+} in Na-Y-B-silicate glass measured at identical delay (5 ms) for $T = 300$ K after laser excitation (a) and the concentration dependence of the energy migration macroparameter γ_H (b).

The existence of a broad transition range may cause substantial errors in determining the order S and the remaining macroparameters in the case of large t_1 and a narrow dynamic range (with respect to t and I) for investigating spectral migration, which is particularly dangerous for specimens with high active ion concentrations.

The migration kinetics on coordinates lg$I(t)$ $-$ $t^{3/10}$ are shown in Figure 32b. It is clear that on these coordinates for $t > t_1$ the curve is accurately linearized which confirms the quadrupole-quadrupole Eu^{3+} $-$ Eu^{3+} interaction mechanism, while the slope of the line gives a value for the macroparameter $\gamma_H = 5.0$ s$^{-3/10}$ and the average energy migration rate $W_H = 2.1 \cdot 10^3$ s^{-1}.

The concentration dependence of the spectral migration rate can be established from an analysis of the luminescence spectrum measured at an identical time delay ($t_D = 5$ ms) for specimens with different Eu^{3+} ion concentrations (Figure 33a). The rise in intensity of the broad pedestal at (λ_{max}

= 578.3 nm) relative to the intensity of the narrow spectral component (λ_{max} = 576 nm) reflects the concentration growth of the spectral migration macroparameters λ_H. Normalizing the sum of the areas of the narrow and broad luminescence bands to unity ($I_B(t) + I_N(t) = 1$) we can give the concentration relation as $\lg[I_B(t)/I_N(t) + 1] = f(n)$ (Figure 33b). As we see this relation is linear which accurately corresponds to Expressions (51), (52) for the disordered spectral migration stage with values of the macroparameters C_{DD} and S independent of n:

$$\lg[I_N(t)/I_B(t)+1] = 0.434\gamma_N(n)t_D^{3/10} = 0.434t_D^{3/10}n(4/3)\pi\Gamma(1-3/S)XC_{DD}^{3/0}$$

$$(174)$$

The nonunity value of $1 + I_B(t)/I_N(t)$ for $n \to 0$ is due to the migrationless, intracenter excitation mechanism of the broad pedestal via the vibronic levels of the Eu^{3+} ions which absorb at λ_E (see the first paper in the present volume). The value of the macroparameter $\gamma_H = 5.0$ s$^{-3/10}$ obtained experimentally makes it possible to use Formula (52) to find the value of the microparameter C_{DD} characterizing migrational interaction efficiency in the pair of Eu^{3+} ions: $C_{DD} = 3.8 \cdot 10^{-3}$ nm^{10}/ms.

Returning to an analysis of Figure 31b it is clear that for the most extensively analyzed concentrations ($c = 12$ mass.% Eu^{3+}) that in the ordered section $t < t_1$ the kinetics decay exponentially at a rate $W_{max} = 5.29 \cdot 10^3$ s^{-1} which exceeds by more than an order of magnitude the average migration rate W_H. Substituting the data on the maximum migration rate W_{max}, the impurity concentration, the multipolarity S and the efficiency C_{DD} into Expression (50) we obtain the last unknown macroparameter characterizing the minimum possible distance of the active Eu^{3+} ions in the medium R_{min}. For the case of Eu^{3+} ions in Na-Y-B-silicate glass this figure was 4.7 Å. Determining the disordered and ordered decay stages as well as R_{min} and t_1 makes it possible to estimate the fraction of Eu^{3+} ions having a spatially (distance) ordered set of nearest neighbors ($R_{ij} = R_{min}$)$P_N = 0.52$ and the fraction of ions belonging to the disordered set $P_D = 0.48$ (having no single neighbor at $R_{ij} = R_{min}$). The high multipole order ($S = 10$) of the test migration among Eu^{3+} ions in glass, as we see from the estimate $t_2 = 213W_H^{-1} = 1$ s, makes it possible to expect simple detection of the diffusion migration stage in this system.

As demonstrated above the true disordered spectral migration range for which the chaotic distribution of the active ions is responsible and which is described by the disordered Förster kinetics makes it possible to determine the macroparameters S and C_{DD}, is bounded on the time scale by the two boundary times t_1 and t_2 (ordered and diffusion stages).

Neglecting the complex, three-stage nature of the decay kinetics the attempt to use simply Förster kinetics for description may result in significant errors in interpreting results, particularly when investigating high-concentration media where the area between t_1 and t_2 may account for only an insignificant region in the decay kinetics. This may cast some doubt on the treatment of the spectral migration energy results obtained by the authors of Refs. [106, 110, 158-160] for calcium metaphosphate with europium. The large values of R_{min} for the rare earth phosphates (R_{min} = 5.6 Å [87]) and small values of C_{DD} (Eu^{3+} - Eu^{3+}) (particularly at low temperatures $T \leqslant 77$ K) indicate a large boundary time t_1. The authors' use [106, 110, 158-160] of specimens with very high Eu_2O_3 concentrations (20-100 mol.%) and the limited measurement time range more closely corresponds to the initial, ordered migration range and the transition ($t \sim t_1$) migration range rather than its second, disordered stage. This explains their diminished multipolarity value of $S = 6$ and the substantial spread of the average migration rates W_H calculated for the same test object neglecting the ordered stage of the process.

We believe that this situation in processing of experimental results is similar to that which existed in the very early stages of kinetic analyses of the energy transfer process [9-12] when the majority of authors attempted to describe the nonexponential Förster kinetics by means of some exponential curve. The authors succeeded in achieving this with some degree of accuracy within a moderate time range although since the Förster decay rate $(3/S)\gamma_H t^{3/S-1}$ depends on the measurement time t the reproducibility of the results and their agreement with theory were poor.

Experimental investigation of the disordered and diffusion stages of migration kinetics for dipole-dipole interaction of Yb^{3+} *ions* [152, 153, 161, 162]. We carried out studies of high-temperature spectral energy migration among Yb^{3+} ions in Ba-Al-phosphate glass of concentrations $n_1 = 1.1 \cdot 10^{20}$ cm^{-3} and $n_2 = 2.9 \cdot 10^{20}$ cm^{-3} in order to experimentally detect and analyze the diffusion-like excitation migration stage in a disordered, highly-dilute medium. A nanosecond LiF crystal/color center laser was used for selective laser excitation (see the first paper in the present volume [152]). Yb^{3+}-activated glasses are preferred for such research since, first, the optical transitions in Yb^{3+} are stronger than those in Eu^{3+} and, second, excitation migration among the Yb^{3+} ions manifests a dipole-dipole nature (see Section 5.1), while the interaction is quadrupole-quadrupole among the Eu^{3+} ions. It is therefore possible to expect an earlier onset of the diffusion stage of the migration process since the boundary time t_2 for dipole-dipole interaction is minimized and is estimated at $9/W_H$.

Unlike the previous low-temperature spectrally-selective energy transfer process among Yb^{3+} ions involving the emission of phonons investigated earlier (see Section 5.1) for this object a temperature rise to $T = 80$ K ($kT \sim \Delta$; ΔE) will result in a virtual total loss of selectivity in the spectral energy migration process. The sudden change in migration type may be attributed to the thermal stimulation of both the Stokes and anti-Stokes nonresonant interaction mechanisms (involving the phonons and the lower Stark sublevels 1 and 1′) as well as the enhancement of the new nonresonant Yb^{3+} ion interaction channel for thermal population of the excited Stark components of the ground (2, 3, 4) and metastable (2′, 3′) levels. The significantly higher homogeneous broadening level and the low degree of energy correlation between the new, thermally-stimulated electron resonances compared to the primary 1-1′ transition where excitation and recording of the spectra occur produce a frequency-independent interaction mechanism of the different optical centers comprising the inhomogeneously-broadened contour.

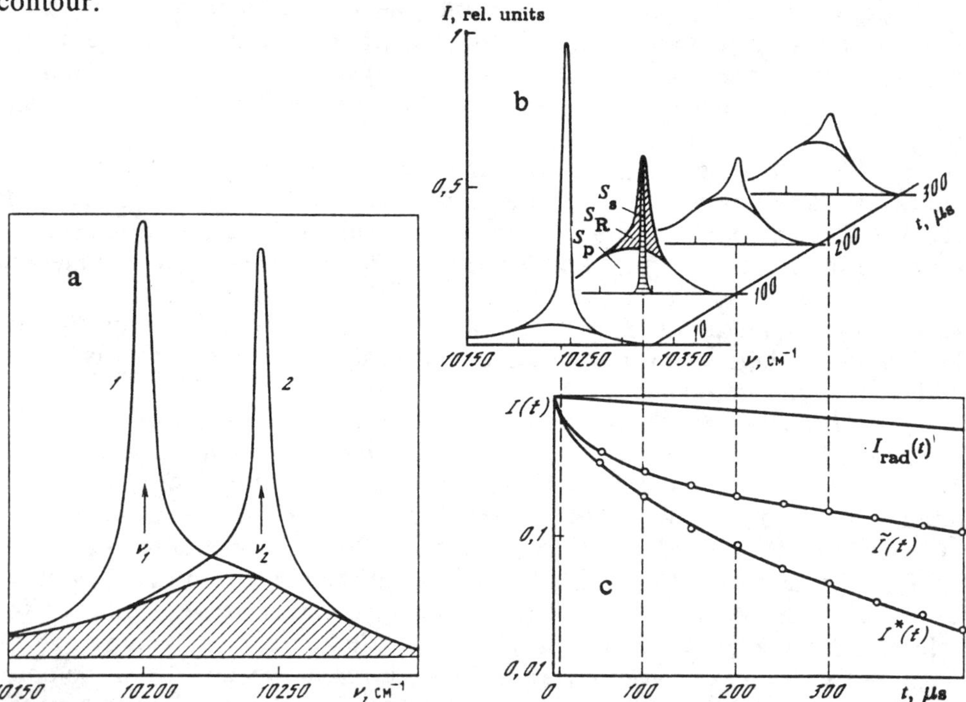

Fig. 34. Selective luminescence spectra of Yb^{3+} ions in Ba-Al-phosphate glass ($n = 2.9 \cdot 10^{20}$ cm^{-3}) for two monochromatic laser excitation frequencies (a) and the spectral-time energy migration dynamics over the inhomogeneously-broadened $^2F_{5/2}$—$^2F_{7/2}$ band (b, c).
a: 1 - $\nu_E = 10{,}200$ cm^{-1}, $t_D = 100$ μs; 2 - 10,242 cm^{-1}, 50 μs.

Figure 34a represents confirmation of the frequency-independent nature of energy migration at $T = 80$ K; this figure shows the two luminescence spectra of the Yb^{3+} ions recorded at the same temperature yet different excitation frequencies: the long-wavelength wing ($\nu_E = \nu_1 = 10{,}200$ cm^{-1})

and the short-wavelength wing ($v_E = v_2 = 10{,}242$ cm^{-1}) of the inhomogeneously-broadened absorption contour. An analysis of the shape of the inhomogeneously-broadened "acceptor" pedestal (the hatched region of the spectrum in Figure 34a) arising in the migration process has demonstrated that it is independent of the laser excitation frequency, indicating an identical energy transfer rate from centers excited by the laser with $v = v_E$ to other optical centers with different frequencies $v(C_{DD} \neq f(v_E - v))$.

The significantly higher radiative probabilities for the optical transitions involved in the interaction in the Yb^{3+} ion case demonstrates the significantly faster dynamics of the high-temperature spectral migration process compared to the case of Eu^{3+} ions even at substantially lower concentrations.

We employed the same experimental and calculation techniques to investigate spectral migration among Yb^{3+} ions as in the Eu^{3+} ion case (see Figure 31).

Figure 34b, c shows results from using such an approach to investigate reversible spectral-spatial excitation migration among Yb^{3+} ions in Ba-Al-phosphate glass ($n = 2.9 \cdot 10^{20}$ cm^{-3}, $T = 80$ K). Here selective laser excitation was used at the center of the $^2F_{7/2}(1)$—$^2F_{5/2}(1)$ band at an excitation frequency $v_E = 10{,}242$ cm^{-1}. It is clear that excitation migration causes the luminescence spectrum of the inhomogeneous set of interacting centers to decay into two components (see Figure 34b): a narrow resonance peak (of area S_R) and a broad inhomogeneously-broadened pedestal (of area S_p). In order to determine the decay kinetics of the initially-excited centers we carried out a measurement of the luminescence decay kinetics $\tilde{I}(t)$ by selective recording at the selective excitation frequency. The spectrometer transmission contour (its area S_s) is shown in Figure 34b. The contribution of the pedestal to the kinetics $\tilde{I}(t)$ is sufficient over rather long times and in order to eliminate this effect the $\tilde{I}(t)$ kinetic curve was corrected every 50 μs by multiplication by the coefficient $k(t_D)/k(0)$:

$$I^*(t) = \tilde{I}(t)\, k(t_3)/k(0), \tag{175}$$

where $k(t_D)$ is the dimensionless correction coefficient calculated by the time-resolved spectra as the ratio of the areas:

$$k(t_D) = S_R(t_D)/S_s(t_D). \tag{176}$$

A difference between the curves $I^*(t)$ and $\tilde{I}(t)$ is clearly evident in Figure 34c and becomes substantial over long time periods. The spontaneous radiative excitation decay channel was eliminated by dividing the kinetic curve $I^*(t)$ by the kinetic radiative decay curve $I_{rad}(t)$ (see Figure 34b) which was also measured by the double spectral selection technique at $T = 4.2$ K for a specimen with a rather low activator concentration ($n = 2.9 \cdot 10^{19}$ cm^{-3})

where migration was reduced:

$$I(t) = I^*(t)/I_{rad}(t). \tag{177}$$

Therefore certain donor luminescence kinetic curves for a specimen with a Yb^{3+} concentration of $n = 1.1 \cdot 10^{20}$ cm^{-3} are given in Figure 35.

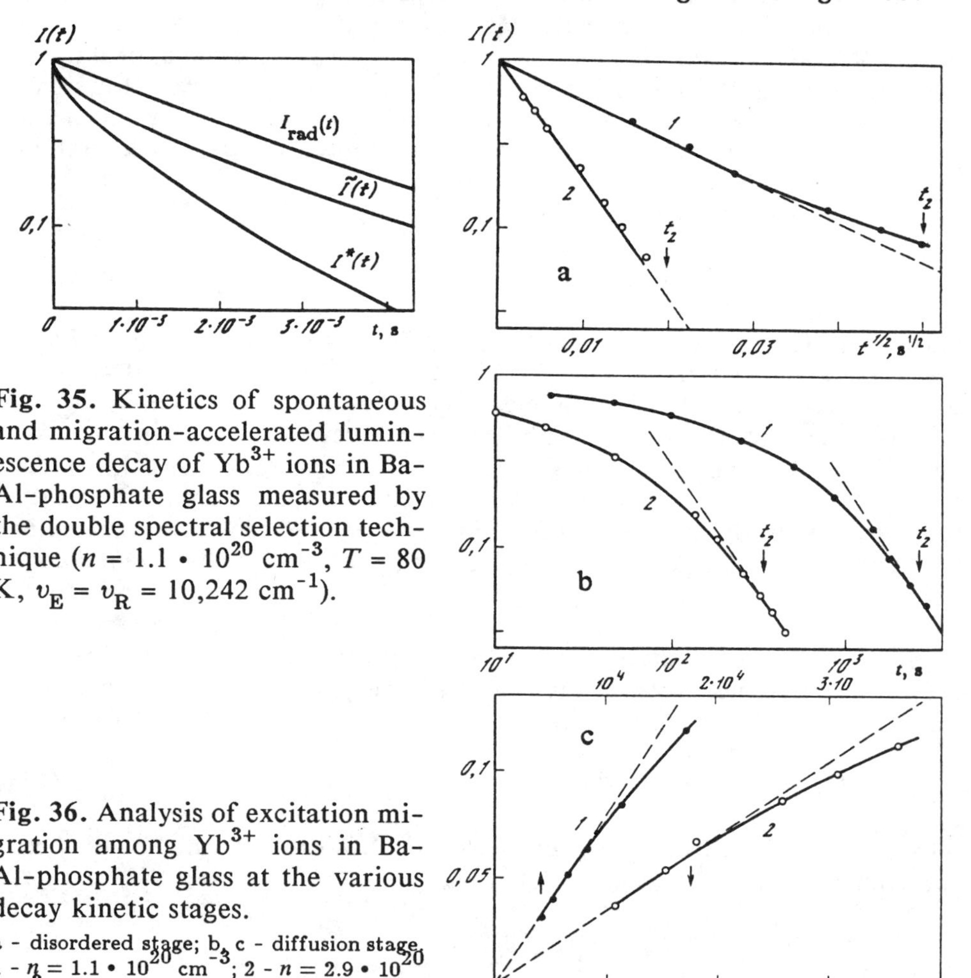

Fig. 35. Kinetics of spontaneous and migration-accelerated luminescence decay of Yb^{3+} ions in Ba-Al-phosphate glass measured by the double spectral selection technique ($n = 1.1 \cdot 10^{20}$ cm^{-3}, $T = 80$ K, $v_E = v_R = 10,242$ cm^{-1}).

Fig. 36. Analysis of excitation migration among Yb^{3+} ions in Ba-Al-phosphate glass at the various decay kinetic stages.

a - disordered stage; b, c - diffusion stage; 1 - $n = 1.1 \cdot 10^{20}$ cm^{-3}; 2 - $n = 2.9 \cdot 10^{20}$ cm^{-3}

As demonstrated above plotting the kinetic curves of donor luminescence on coordinates lg $I(t) = t^{3/S}$ permits identification of the ion-ion transfer mechanism and makes it possible to determine the parametric relations for the disordered stage by means of Expression (52). The rectification of the kinetic curves on these coordinates for $S = 6$ (Figure 36a) confirms the dipole-dipole nature of interaction between the Yb^{3+} ions in Ba–Al–phosphate glass and makes it possible for this stage to find the excitation macroparameters γ_H from the slopes:

$$\text{for } n_1 = 1.1 \cdot 10^{20} \text{ cm}^{-3} \quad \gamma_H(n_1) = 55 \text{ s}^{-1/2}$$

$$\text{for } n_2 = 2.9 \cdot 10^{20} \text{ cm}^{-3} \quad \gamma_H(n_2) = 152 \text{ s}^{-1/2}.$$

The experimental error in determining γ_H which was estimated by the spread of the experimental points was approximately 5%.

We note that the ratios of the ion concentrations ($n_2/n_1 = 2.64$) and the dipole-dipole interaction parameters ($\gamma_H(n_2)/\gamma_H(n_1) = 2.76$) were nearly identical, suggesting a linear (within error) relation $\gamma_H(n)$.

Knowledge of γ_H makes it possible, using the expressions $W = \gamma^{S/3}$, to determine the average transfer rate $W_H(n)$ for the nonstationary stage:

$$W_H(n_1) = 3 \cdot 10^3 \text{ s}^{-1}, \, W_H(n_2) = 2.3 \cdot 10^4 \text{ s}^{-1}.$$

An increase in concentration by a factor of 2.64 will result in a near order of magnitude increase in the migration rate. This in turn sharply reduces the duration of the nonstationary stage.

The values of the macroparameter γ_H and the multipole order S can be used with Expression (52) to calculate the most important macroparameter reflecting the efficiency of dipole-dipole high-temperature reversible interaction between Yb^{3+} ions in Ba–Al–phosphate glass: $C_{DD}(Yb^{3+} - Yb^{3+}) = 9.0 \text{ nm}^6/\text{ms}$.

A comparison of the macro- and microparameter values obtained at $T = 80$ K which are common to all Yb^{3+} optical centers to their low-temperature analog dispersion curves (see Section 5.1) reveals a significant increase in interaction efficiency and spectral migration rates with increasing temperature. We note that the value $C_{DD}(Yb^{3+} - Yb^{3+})$ in phosphate glass makes it possible to calculate the minimum possible energy transfer probabilities in this medium over the distance R_{min}. The minimum distance between Yb^{3+} ions in accordance with Ref. [6] may lie in the 4-6 Å range. For an average $R_{min} = 5$ Å the elementary transfer rate $W_{max} = 5.8 \cdot 10^5 \text{ s}^{-1}$. This is a fundamental quantity for estimating the efficiency of excitation energy migration among Yb^{3+} ions to the Er^{3+} laser ions in erbium laser phosphate glass operating as a radiation converter ($\lambda_{pump} = 1.06 \, \mu\text{m}$, $\lambda_{con} = 1.54 \, \mu\text{m}$).

Figure 36b shows kinetic curves replotted on coordinates $\lg I(t) - \lg t$, which makes it possible to analyze their behavior at rather late times. On these coordinates the "root" law becomes the power-law relation

$$\lg I(t) = -\frac{1}{2,3} \gamma_H \cdot 10^{\frac{1}{2}\lg t} , \tag{178}$$

while this is replaced by a linear relation with a slope of 3/2 indicating diffusion behavior of the kinetics.

This fact represents the first experimental confirmation of the existence of a diffusion excitation migration stage in a low-concentration disordered solid. It is important to point out that the direct experimental proof of the existence of excitation diffusion in a disordered medium refutes previous doubts regarding the validity of describing the long-term asymptotics of the migration process by means of the expression $I(t) = (W_{diff}t)^{-3/2}$ which on the coordinates outlined above appears as

$$\lg I(t) = -\frac{3}{2}\lg(W_{diff}t). \tag{179}$$

The boundary time for the onset of the diffusion stage is the time where Expression (179) begins to be valid. It is evident from Figure 36b that the boundary time t_2 is approximately equal to $2.8 \cdot 10^{-3}$ and $3.8 \cdot 10^{-4}$ s, respectively, for $n_1 = 1.1 \cdot 10^{20}$ cm^{-3} and $n_2 = 2.9 \cdot 10^{20}$ cm^{-3}, which is in good agreement with the values calculated from previously-derived expressions. Indeed, estimates employing (171a) provide the following boundary times for n_1 and n_2, respectively: $3.0 \cdot 10^{-3}$ and $3.9 \cdot 10^{-4}$ s. This fact also supports the validity of the estimate based on measurement of the macro-parameter γ_H; the determination of this parameter from the nonstationary kinetic stage remains a rather popular technique today.

Additional confirmation of the existence of the diffusion stage of disordered migration is the rectification of migration kinetics $I(t)$ in the function of the argument $t^{-3/2}$ and its vanishing as $t \to \infty$ (see Figure 36c). It is clear from (41) that the slope of the linear section of this relation as $t^{-3/2} \to 0$ provides quantitative values for the rate of the process in the diffusion stage W_{diff}. For a specimen of concentration $n_1 = 1.1 \cdot 10^{20}$ cm^{-3}, $W_{diff}(n_1) = 2.5 \cdot 10^{3}$ s^{-1}, while for a specimen of concentration $n_2 = 2.9 \cdot 10^{20}$ cm^{-3}, $W_{diff}(n_2) = 2 \cdot 10^{4}$ s^{-1}.

It is evident that the migration rate in the diffusion stage, as is the case on the nonstationary section, has a square-law dependence on the activator ion concentration, while the number values of the average migration rate determined from the nonstationary (W_H) and the diffusion (W_{diff}) stages of the process are similar.

Determination of the diffusion coefficients and their dependences on concentration also occupy an central role in investigating the long-term asymptotics of these processes. We have already demonstrated the validity of describing the long-term asymptotics of the migration process in a disordered particle ensemble using Expression (41) which was derived exactly for the case of an ordered lattice. This makes it possible to assume that the relation between the average rate W_{diff} and the diffusion coefficient is common to ordered and disordered media and Expressions (42) can be used. The excitation diffusion coefficients in a disordered medium calculated using this expressions for our case were:

for n_1 $D_H = 0.9 \cdot 10^{-11}$ cm^2/s,

for n_2 $D_H = 3.6 \cdot 10^{-11}$ cm^2/s.

The diffusion coefficients measured directly from experiment for the two ion-activator concentrations correspond to a relation $D_H \sim n^{4/3}$ which represents the first experimental confirmation of an irrational power-law dependence of the diffusion coefficient on the concentration of randomly distributed centers.

So far we have discussed the results obtained from selective excitation of luminescence at the center of the broadband inhomogeneous $^2F_{5/2}(1)$ $-^2F_{7/2}(1)$ contour ($v_E = v_R = 10{,}242$ cm^{-1}). It would also be interesting to investigate the dynamics of the deactivation of the initially-excited set of centers by varying the excitation and recording frequencies over the entire absorption contour of the Yb^{3+} ions.

The technique discussed above was also used for other frequencies on the low-frequency and high-frequency wings of the contour ($v_E = v_R$), specifically 10,200 and 10,257 cm^{-1} for this purpose. The luminescence spectra with different recording delays were recorded for each of these frequencies and kinetic luminescence decay curves were plotted for identical selective excitation and selective luminescence recording frequencies. The experimental curves were corrected in accordance with Expressions (175) and (176). The kinetic decay curves of the initially-excited centers were obtained in this manner; these curves are shown in Figure 37a.

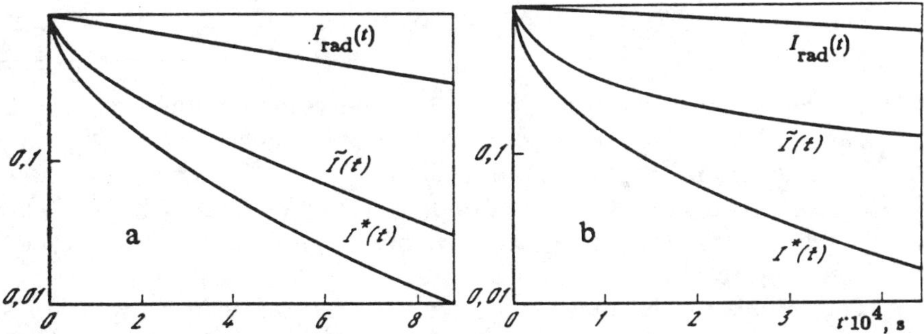

Fig. 37. Kinetics of spontaneous and migration-accelerated luminescence decay of Yb^{3+} ions in Ba-Al-phosphate glass using the double spectral selection technique for $n = 2.9 \cdot 10^{20}$ cm^{-3}, $T = 80$ K and $v_E = v_R = 10{,}200$ (a) and 10,257 cm^{-1} (b).

These curves are replotted on coordinates $(-\lg I(t)) - \lg t$ in Figure 38a. From the functional viewpoint the picture is identical to the case examined previously (see Figure 36b): a law similar to the Förster law describing the nonstationary stage, and on these coordinates it yields a power-law relation (Expression (178)) which gradually becomes a linear relation with a slope of 3/2. It is clear from this figure that an increase in v_E from 10,200 to 10,257 cm^{-1} results in an earlier onset of the diffusion stage. The values of the boundary time for the onset of the diffusion stage t_2^O obtained from these data are given below:

$$v_E = v_R, \text{ cm}^{-1} \quad 10,200 \qquad 10,242 \qquad 10,257$$

$$t_2^O, \text{ s} \qquad 6.5 \cdot 10^{-4} \quad 3.9 \cdot 10^{-4} \quad 2.5 \cdot 10^{-4}$$

It would be natural to expect that such a change in boundary time t_2 as a function of frequency is correlated with the values of the excitation migration macroparameters of the nonstationary and diffusion stages. In order to determine this we plot kinetic luminescence decay curves on the coordinates $\lg I(t) - t^{1/2}$ for $v_E = v_R = 10,200$ and $10,257$ cm^{-1} (Figure 38b) and coordinates $I(t) - t^{-3/2}$ (Figure 38d).

Table 4

Macro- and microparameters for different
laser excitation frequencies

v, cm^{-1}	γ_H, s$^{-1/2}$	$W_H \cdot 10^{-4}$, s^{-1}	c_{DD}, nm^6/µs	t_2^T, µs
10 200	114	1.3	6	690
10 242	152	2.3	9	380
10 257	185	3.4	15	270

We can determine from Figure 38b the values of the macroparameters γ_H, as well as their corresponding average transfer rates W_H and the transfer efficiency macroparameters $C_{DD}(\text{Yb}^{3+} - \text{Yb}^{3+})$ which are shown in Table 4. Also given here are the boundary times estimated by the expressions $t_2^T = 9W_H^{-1}$ which are in good agreement with the values of t_2^O determined from Figure 38a.

The values of the macroparameter of the diffusion migration stage W_{diff} were determined by means of Figure 38c: $1.2 \cdot 10^4$ s^{-1}, for $v_E = v_R = 10,200$ cm^{-1} ($D_N = 2.2 \cdot 10^{-11}$ cm^6/s, respectively), and $2.7 \cdot 10^4$ s^{-1} for $v_E = v_R = 10,257$ cm^{-1} ($D_N = 4.9 \cdot 10^{-11}$ cm^6/s, respectively).

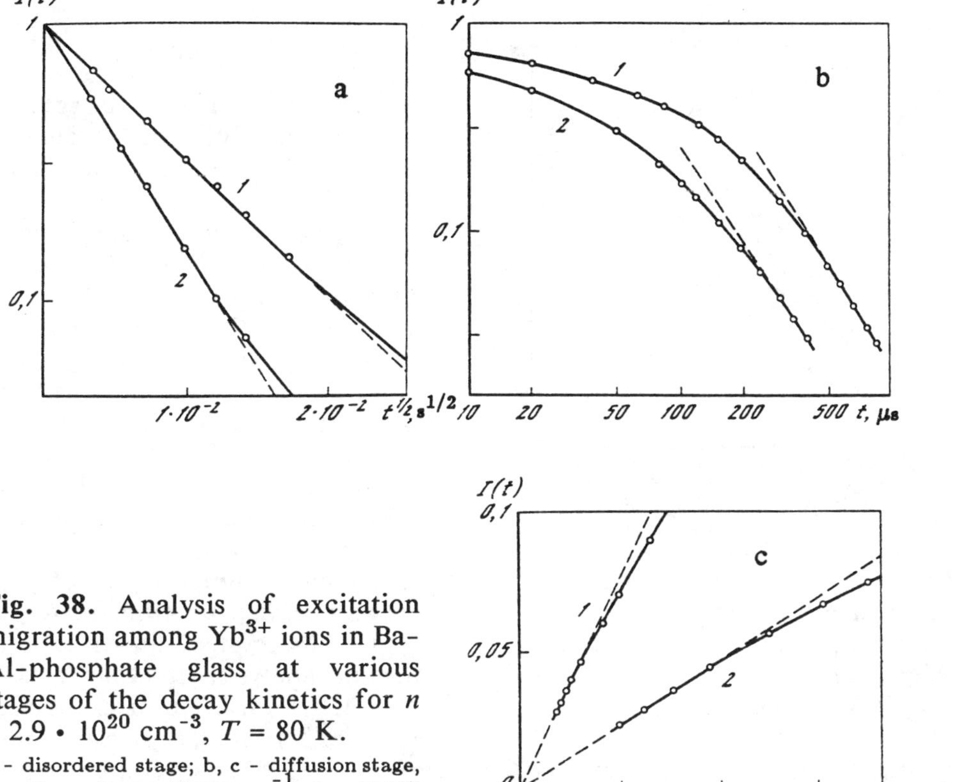

Fig. 38. Analysis of excitation migration among Yb^{3+} ions in Ba-Al-phosphate glass at various stages of the decay kinetics for n = 2.9 • 10^{20} cm^{-3}, T = 80 K.

a - disordered stage; b, c - diffusion stage, 1 - $v_E = v_R = 10,200$ cm^{-1}, 2 - 10,257 cm^{-1}

The average migration rates determined directly from experiment for the nonstationary and diffusion stages of the process are given in Figure 39. The figure reveals that the migration rate in the nonstationary stage W_H, as in the diffusion stage W_{diff}, grows as $v_E = v_R$ goes from the low-frequency wing to the high-frequency wing which is in good agreement with the dispersion of the radiative transitions within the inhomogeneous contour. This makes it possible to treat the increase in the average migration rate as an increase in the microparameter C_{DD} (see Table 4) which is directly proportional to the radiative transition probability of the donor: $C_{DD}(v_E) \sim A(v_E)$.

The independent determination of the migration rates W_{diff} and W_H of the disordered diffusion coefficients D_d and the elementary interaction microefficiency C_{DD} from various stages of the migration process allows us to establish a quantitative relation between these by estimating the numerical coefficient α in Expression (74) for dipole-dipole interaction:

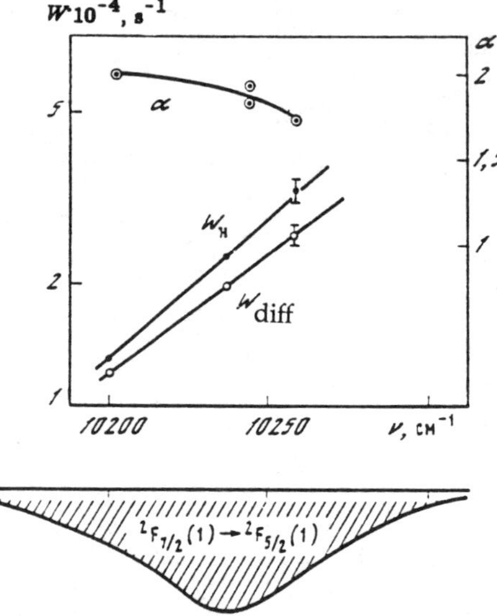

Fig. 39. Frequency dependences of the macroparameters of high-temperature reversible excitation migration at the $^2F_{7/2}(1)-{}^2F_{5/2}(1)$ transition of Yb^{3+} ions.

The values of α for the various selective laser excitation wavelengths are given in Figure 39.

It is clear that the experimentally-determined values of the parameter α vary monotonically and adopt values lying between those predicted by theory (α = 2.2-3.5) and those obtained by computer modeling (α = 1.55 ± 0.7). This may be due to the varying degree to which the theoretical models take account of energy back transfer and possible deviation from complete symmetry of forward and back transfer ($W_0 \neq W_{i0}$) in the system under experimental analysis. In this case a monotonic variation of $\alpha(\nu_E)$ may be an experimental manifestation of this asymmetry or may reflect the specific nature of energy migration in the system of centers having a dispersion of the radiative probabilities $A(\nu_E)$.

Two experimental studies appeared in the years following the publication of Refs. [1] which were the first to detect and investigate the diffusion stage of the migration kinetics in a disordered ensemble of particles and establish the relation $D \sim n^{4/3}$.

Ref. [163] measured the picosecond kinetics of the expansion of nonstationary excitations of dye molecules in solution. The authors of Ref. [163] also found the diffuse excitation expansion mode (delocalization), confirmed the relation $D \sim n^{4/3}$ and identified, as we have, reduced experimental diffusion coefficients compared to theoretical values. We found a parameter α = 1.36 based on these data which, however, unlike our case, was based on an estimate rather than an independent measurement of C_{DD}. Such a low value of α (or D) compared to the theoretical value is attributed by the authors to the effect of the finite molecular dimensions (inaccuracy). However as we have seen from statistical modeling results R_{min} has the strongest effect as $n \rightarrow n_{max}$ ($c \rightarrow 1$), and will either leave unchanged or will increase rather than decrease the diffusion coefficients D and α. In order to compare the results of Ref. [163] to our data it is necessary to account for the orientational effect when analyzing the energy migration process between molecules having a specific dipole moment.

As noted in Ref. [163] orientation averaging reduces the diffusion coefficients and α by a factor of 1.26. The value of the coefficient $\alpha = 1.36 \cdot 1.26 = 1.71$ obtained neglecting the orientation effects more closely correlates with our case of the interaction of Kramers Yb^{3+} ions whose dipole moment more closely corresponds to a spherically-symmetrical rather than a linear oscillator [164, 166]. The value $\alpha = 1.71$ calculated in this manner for an essentially different system and for another range of migration rates correlates reasonably well with our experimental values of $\alpha = 1.74$-2.02 and the values $\alpha = 1.55 \pm 0.7$ obtained by computer statistical modeling.

While using our approach and providing an unsubstantiated critique of it the authors of Ref. [157] made what we believe is an important attempt to increase intensity in measuring migration kinetics by altering the system to some degree (using a LaP_5O_{14} crystal containing Yb^{3+} ions). However the authors did not achieve (or did not detect) the diffusion stage of migration yet provided a complete confirmation of the theory [122] (expression (102)) predicting a strong overvibration in the dilute disordered system. As we have demonstrated above the conclusions of this theory which yield an important result as $t \to \infty$, $(I(t) \sim t^{-3/2})$, substantially diverge (>35%) from solution (51), statistical modeling result and experimental data of a variety of authors [28, 95, 112, 125, 156, 161] as the theory does not properly take account of the transfer reversibility at an intensity $I(t) \sim 0.1 \, I(0)$.

Computer applications to data processing does not eliminate possible methodological errors. Thus, differences in the inhomogeneously-broadened luminescence spectra with both small and large delays demonstrate that setting the spontaneous decay kinetics equal with different v_E is not valid; this is particularly important under the measurement conditions of Ref. [157] over times $t \gg A_{sp}^{-1}W_H^{-1}$. This may result in incorrect normalization and significant distortions to migration kinetics in the later stage (through type (102) overvibrations). This is also supported by the unusually low accuracy in the determination of the migration rate in the initial (Hüber) stage ($W_H = 900 \text{ s}^{-1} \pm 20\%$) at a high signal intensity at the same time that the late stage where the intensity is two orders of magnitude lower the accuracy was for some reason three times higher. We also note that although our estimated coefficient $\alpha = 3.11$ based on the data from Ref. [157] is lower than that predicted by theory [122] ($\alpha = 3.46$) which neglects the reversibility in the diffusion coefficient, is nonetheless higher than all values of α obtained both experimentally and theoretically as well as by statistical techniques.

6. Experimental Investigation of the Kinetics of Static Donor-Acceptor (Sm^{3+}-Sm^{3+}) Quenching Energy Transfer in a Disordered Matrix

We have considered experimental results from an investigation of energy migration among the metastable levels of donors in the absence of acceptors (quenchers) responsible for the nonradiative absolute decay of excitations.

In this section we consider experimental research on another limiting case: direct quenching energy transfer under conditions where donor-donor migration is absent, i.e., when excitation prior to donor luminescence of nonradiative transfer to the acceptor remains fixed.

The problem of establishing the elementary donor-acceptor quenching interaction mechanisms (their multipole order and microefficiency) based on the kinetics of energy transfer and the development of spectral-kinetic techniques for analyzing the spatial distribution of active impurities in disordered solids, as discussed in Section 2, dictate a number of the requirements for selecting the model active medium: $A + \bar{W} \ll \gamma_F/t^{S-3/S} \ll W_{max}$. Above all it is necessary to have a high direct ion-ion quenching interaction probability ($W_{max}(C_{DA}) \gg A$) and a low migrational donor-donor interaction probability ($W(C_{DD}) \ll A$). These conditions do not hold for all rare earth ions.

Hence, cross-relaxational self-quenching is lacking in the majority of metastable levels of the Yb^{3+}, Tm^{3+}, Er^{3+}, Ho^{3+} ions and where it does occur it is accompanied by rapid migration rates due to the spin-allowed optical transitions. Such a situation is also observed for the Nd^{3+} ion which is more interesting from the practical viewpoint; here the rates W_{max} and $\bar{W}$ are often of the same order of magnitude [15, 85]. On the other hand, the migration processes are severely attenuated for Eu^{3+} and Tb^{3+} ions due to the optically-forbidden transitions, although self-quenching is quite low for these as well and can be expected to exist only for the upper metastable levels associated with the lower intracenter nonradiative transitions. We therefore selected the Sm^{3+} ion as the model ion for self-quenching studies since the $^4G_{5/2}$ metastable level of the Sm^{3+} ion makes it possible to excite this ion by a tunable Rhodamine 6Z dye laser. The fact that the $^4G_{5/2}-^6H_{5/2}$ optical transition is forbidden eliminates the effect of migrational interaction in this case at the same time that concentration quenching is rather strongly developed [86].

Li-La-Sm-phosphate glass was used as the medium to investigate the features of the spatial distribution of active ions. This matrix is a highly efficient laser material and has anomalously-weak concentration quenching when Sm^{3+} ions are replaced by Nd^{3+} ions [85].

The tests were carried out on Li–La-phosphate glass specimens of variable Sm^{3+} ion concentration (from $0.15 \cdot 10^{20}$ to $23 \cdot 10^{20}$ cm^{-3}), which were introduced by substitution of La ions. The Sm^{3+} ion concentration was monitored by multiprobe analysis on the "Camebax" instrument and by comparing the absorption spectra for samples with different Sm^{3+} ion concentrations. The luminescence decay kinetics were investigated over a broad dynamic intensity range using a technique based on the virtual instantaneous (10^{-8} s) excitation of Sm^{3+} ions by a tunable dye laser (Rhodamine 6Z) followed by photographing of individual sections of the decay kinetic curves from a S1-70 oscilloscope. An MDR-2 monochromator was used for pretuning of the Sm^{3+} ions to luminescence; the luminescence was then recorded by an FEU-70 photomultiplier. The oscilloscope had a good amplitude-frequency response and a beam sweep linearity of better than 3%. "Instantaneous" laser excitation of luminescence is fundamental here since this is the only case where the spatial distribution of the excited ions will

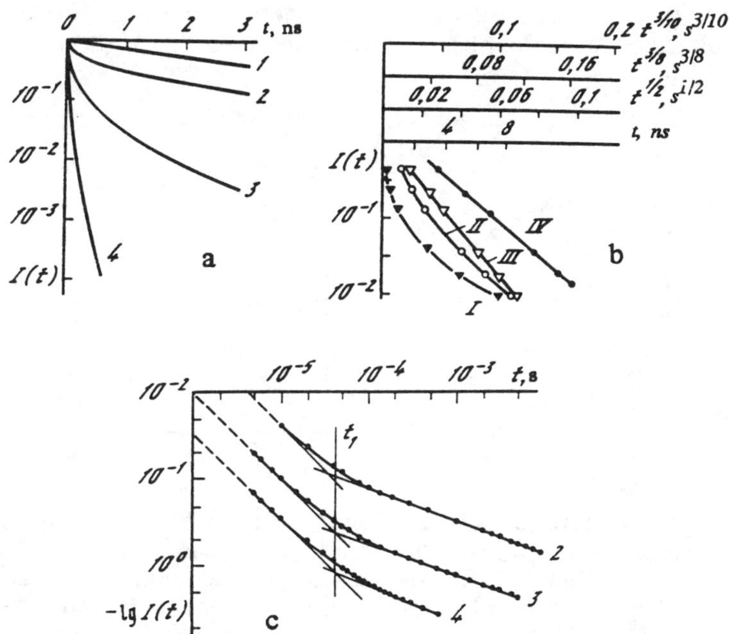

Fig. 40. Kinetics of luminescence decay (a) and nonradiative losses (b, c) from the $^4G_{5/2}$ level for various Sm^{3+} ion concentrations n in glass. a, c: 1 – n = 0.15 • 10^{20}, 2 – 2.5 • 10^{20}, 3 – 9.2 • 10^{20}, 4 – 23 • 10^{20} cm^{-3}; b: I – $f(t)$, II – $f(t^{1/2})$, S = 6, III – $f(t^{3/8})$, S = 8, IV – $f(t^{3/10})$, S = 10.

reflect their spatial distribution in the ground state. This condition must underlie the derivation of Expressions (25), (29) and (30) and makes the observed kinetics more valuable with respect to the information they contain.

The luminescence decay kinetics $I_L(t)$ of the $^4G_{5/2}$ level of Sm^{3+} ions were measured over more than a four order of magnitude intensity range. It is clear from Figure 40a that an increase in the Sm^{3+} ion concentration in glass above $0.15 \cdot 10^{20}$ cm^{-3} results in a sudden enhancement in luminescence quenching manifested as a nonexponential character to the decay kinetics. The better than 10% agreement in quenching times measured in the final exponential decay stages for three Sm^{3+} ions ($0.15 \cdot 10^{20}$, $1.5 \cdot 10^{20}$, $2.5 \cdot 10^{20}$ cm^{-3}) confirms the poor efficiency of the energy migration processes over the metastable $^4G_{5/2}$ level of Sm^{3+} ions ($\overline{W} \to 0$), and permitting analysis of the static nonradiative Sm^{3+} - Sm^{3+} energy transfer mechanism only.

Figure 40b demonstrates the nonradiative transfer kinetics lg $I(t)$ = lg $I_L(t)$ + 0.434 t/τ_0 of a specimen of concentration $9.2 \cdot 10^{20}$ cm^{-3} on different time coordinates. It is clear that linearization of the decay kinetics corresponds to high multipole orders of the process: $S = 8$ and 10. An analogous behavior of the kinetics is observed for specimens with different Sm^{3+} ion concentrations.

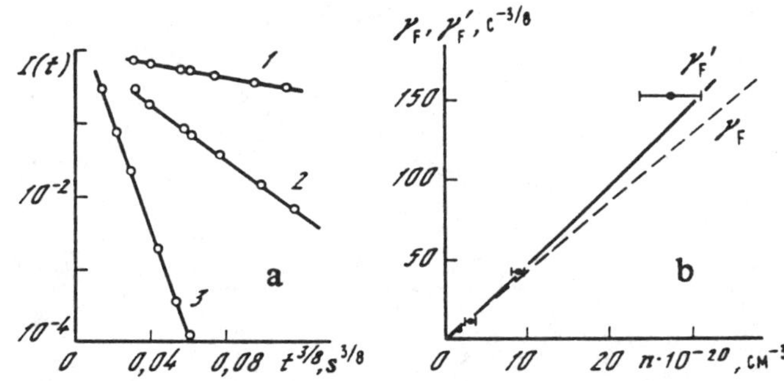

Fig. 41. Concentration quenching of Sm^{3+} ions in Li–La–phosphate glass

a - kinetics of nonradiative losses for various Sm^{3+} ion concentrations: $1 - n - 2.5 \cdot 10^{20}$, $2 - 9.2 \cdot 10^{20}$, $3 - 23 \cdot 10^{20}$ cm^{-3}; b - concentration dependences of the microparameters γ_F and γ_F' (circles: experiment; the spread of the experimental concentration values represents the difference between results from a microprobe analysis and an analysis of the absorption spectra and quantity of Sm^{3+} impurities introduced in glass fabrication).

By plotting the nonradiative loss curve on coordinates lg($-$lg $I(t)$) – lg$t$, it is easy to analyze the different static quenching stages (Figure 40c) The slope (tg φ) of the relations lg($-$lg $I(t)$) = f(lg t) determines the order of the parameter t, thereby differentiating ordered quenching from disordered quenching and making it possible to estimate its multipole order for the latter. It is clear from Figure 40c that the onset of the process for $t < t_1$ is characterized by tg $\varphi = 1$ for all concentrations, which corresponds to exponential ordered decay, while in the final stage of the process tg φ is lower which corresponds to disordered quenching. The slope of the kinetic curve for $t > t_1$ for all three concentrations adopts values of 0.39, 0.36, 0.41 close to 3/8, indicating a dipole-quadrupole interaction mechanism ($S = 8$) of the Sm^{3+} ions.

Figure 41a shows the kinetic dependences of nonradiative losses for various concentrations on $t^{3/8}$ (dipole-quadrupole quenching mechanism).

It is clear that the curves are well-linearized on coordinates lg $I(t)$ – $t^{3/8}$, and it is possible to determine the quenching macroparameter $\gamma_F'(n)$ (see Expressions (30), (32)) from the slope of these curves. The concentration relation of quenching efficiency $\gamma_F'(n)$ calculated in this manner is shown in Figure 41b which clearly reveals a linear nature of $\gamma_F'(n)$ through a concentration $9.2 \cdot 10^{20}$ cm^{-3} and a stronger dependence on n at higher concentrations. There are two possible reasons for the deviation from linearity of the $\gamma_F'(n)$ relation as predicted by theory. One is a concentration variation in the interaction microparameter C_{DA} due to the changing spectral properties of the test material due to structural rearrangement of the Sm^{3+} optical centers at high concentrations. We estimated this probability from a comparison of the reduced absorption spectra at the $^6H_{5/2}-{}^6F_{5/2}$, $^6H_{5/2}-{}^6F_{7/2}$, $^6H_{5/2}-{}^6F_{9/2}$, $^6H_{5/2}-{}^6F_{11/2}$, $^6H_{5/2}-{}^4G_{5/2}$ transitions for different Sm^{3+} ion concentrations. The spectra turned out to be identical, which eliminated this explanation of the nonlinearity of $\gamma_F'(n)$.

The second possible explanation for the nonlinearity derives from quenching theory taking account of the higher terms in the expansion of $\gamma_F'(n)$. In Section 2 we found the relation $\gamma_F' = f(c)$ (32) for the case of higher multipole orders and high acceptor concentrations which were not limited by the condition $c \ll 1$ (as is usually the case). Such an expression makes it possible to explain the nonlinearity of the relation $\gamma_F' = f(n)$ detected here if the parameter $c = n/n_{max}$ approaches one for concentrations $n \geqslant 2.3 \cdot 10^{20}$ cm^{-3}. This is possible at moderately high values of n_{max} in the medium when the forbidden volume occupied by each ion and, therefore, the minimum possible distance R_{min} are sufficiently great.

The experimentally-determined values of the macroparameter γ_F' which grows linearly in the low concentration range $n(Sm^{3+}) \leqslant 9.2 \cdot 10^{20}$ cm^{-3}, make it possible by using Formula (31) to find the microparameter C_{DA} which characterizes the efficiency of $Sm^{3+} \rightarrow Sm^{3+}$ quenching interactions. This was equal to 0.23 nm^8/ms. Experimental data obtained at 77 K are essentially identical to data obtained at 300 K, indicating a constant value of the microparameter C_{DA} in this temperature range.

tration quenching by recording their luminescence in different regions of the decay kinetic curve. A sample such monochromatic excitation spectrum for Sm^{3+} ions in the test glass is given in Figure 42b for two delays between the excitation time and the luminescence recording time t_D: $5 \cdot 10^{-6}$ and $3.5 \cdot 10^{-4}$ s.

It is clear from the figure that increasing the delay causes a strong deformation in the excitation spectrum. The relative drop in luminescence intensity observed from excitation to the short wavelength edge of the absorption spectrum at $t_D = 3.5 \cdot 10^{-4}$ s indicates a more strongly-expressed luminescence quenching for these groups of centers (we recall that $A \neq f(\lambda_E)$ and $\tau_0 \neq (\lambda_E)$ for Sm^{3+} ions in glass (see the first paper)). If we know the luminescence decay kinetics for one excitation wavelength $\lambda_E = 5612$ Å (see Figure 40a) and by comparing excitation spectra 1 and 2 in Figure 42b it is possible to calculate the change in magnitude of the quenching parameter γ_F and the microparameter C_{DA} as a function of laser excitation wavelength for the different optical centers. The latter relation is shown in Figure 42c and demonstrates that the ion-ion interaction efficiency increases for centers with higher metastable level energies. The proposed method of investigating quenching substantially simplifies the analysis of ion-ion energy transfer in a system with a continuous set of optical centers.

In analyzing the experimental data on luminescence quenching we have been interested solely in the final, disordered stage of the luminescence decay kinetics at the same time that we have noted the important role of the "ordered" kinetic stage ($0 < t < t_1$). The ordered decay kinetic stage bears direct information on the most important parameters relating to the spatial configuration of the impurity ions; specifically, the minimum possible distance between particles and the fraction of ions P_O in the spatially ordered state.

Figure 43a shows the ordered and transitional luminescence kinetic decay stages for different Sm^{3+} concentrations in the test glass. It is important that the ordered kinetic stage ($0 < t < t_1 = 40$ μs) corresponds to decay of a substantial number of luminescing centers whose decay rate and relative concentration grow with increasing total impurity concentration n (Sm^{3+}) in the glass.

Using data on the interaction multipole order $Sm^{3+} \rightarrow Sm^{3+}$ ($S = 8$) and the interaction microparameter $C_{DA} = 0.23$ nm^8/ms obtained from the analysis of the disordered kinetic stage we can substitute the rates into Expression (28) for the initial disordered luminescence quenching W_{max} (see Figure 43a) for different acceptor concentrations n and find the unknown microparameter R_{min}. Irrespective of the clear difference between the luminescence quenching efficiency $W_{max}(n)$, the calculated minimum distance remained constant ($R_{min} = 5.6$ Å) across nearly the entire ten-fold concentration range of Sm^{3+} (substitution of La^{3+} by Sm^{3+}) indicating that the structure of this series of highly-concentrated glasses remains unchanged.

Knowledge of R_{min} is important in analyzing the quenching of the impurity centers. As we see from Equation (28) the high power of the dependence of quenching probability on R_{min} ($W_{max} \sim R_{min}^{-5}$ for Sm^{3+} ions and $W_{max} \sim R_{min}^{-3}$ for Nd^{3+} ions) suggests a strong enhancement of quenching with diminishing distance R_{min}. One of the fundamental requirements on the materials with anomalously-weak quenching derives from this property. The high value of R_{min} = 5.6 Å for Li-La-Sm-phosphate glass which is close to R_{min} for certain highly-concentrated crystal phosphates (R_{min} = 5.2-6.5 Å [169]) in all probability is one of the most important causes of the anomalously-weak rare-earth ions quenching in this glass.

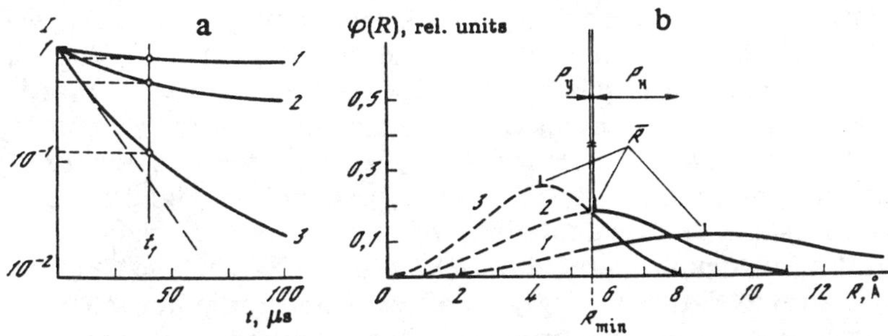

Fig. 43. Initial stage of the luminescence decay kinetics in specimens with a high Sm^{3+} ion concentration (a) and the distribution functions of the ion-ion distances for the nearest neighbors calculated by the point particle model (R_{min} = 0, dashed curves) and taking the finite particle dimensions into account (R_{min} = 5.6 Å, solid curves) (b).
n, cm^{-3}: 1 - 2.5 • 10^{20}, 2 - 9.2 • 10^{20}, 3 - 23 • 10^{20}.

We proposed Expression (34)-(36) to determine the fractions of Sm^{3+} ions in the spatially ordered (P_O) and spatially disordered (P_D) particle groups based on the luminescence decay kinetics (see, for example, Figure 43a). Calculation by these formulae yields the following relative concentrations of P_O and P_D as a function of the total Sm^{3+} ion concentration in the glass:

n, cm^{-3}	2.5 • 10^{20}	9.2 • 10^{20}	2.3 • 10^{21}
P_O	0.15	0.53	0.88
P_D	0.85	0.47	0.12

An increase in the total Sm^{3+} concentration clearly serves to increase the fraction of ions at the minimum possible distance ($R_{ij} = R_{min}$, ordered set) compared to ions having randomly-configured neighbors ($R_{ij} > R_{min}$).

The lack of a strict exponential decay across the entire ordered decay stage does not represent a contradiction even at Sm^{3+} concentrations of $n = 2.3 \cdot 10^{21}$ cm^{-3} where, as clearly indicated by the data presented above, the majority of ions are in the ordered state. In our case "ordering" characterizes only an unchanging ion-ion distance: $R_{ij} = R_{min}$, yet this does not require a fixed number of nearest neighbors which can vary statistically about the mean ac if the particle concentration differs from the "maximum" possible concentration ($n < n_{max}$) given by R_{min} and the maximum coordinational packing number a of the impurity sets for the given medium. The value of R_{min} permits determination of the maximum possible impurity concentration in glass of a given composition for the given coordinational number.

In our case using $R_{min} = 5.6$ Å and the maximum possible coordinational number 12 (densest packing of spherical SmO_8 complexes) we obtain $n_{max} = 8 \cdot 10^{21}$ cm^{-3}. Comparing n_{max} to the highest Sm^{3+} (or La^{3+}) ion concentration in the test specimens ($2.3 \cdot 10^{21}$ cm^{-3}) we see that the glass under test still has a far from maximum active ion concentration n_{max} where all ions will be at identical distances ($R_{min} = 5.6$ Å) with an equal number of neighbors, and the kinetics become strictly exponential along the entire time scale. At the same time calculations of the number of aggregate centers (pairs, triples, quadruples, etc.) at $c = 0.29$ ($n = 2.3 \cdot 10^{21}$ cm^{-3}) indicate that ordering is already rather strongly manifested and the point ion approximation becomes unsuitable. As discussed above this must be accounted for in analyzing the concentration quenching relations in the exponential (Förster) region. It was not, however, possible for us to take this into account previously due to a lack of data on n_{max} and c. Substituting the estimate for n_{max} into Expression (32) it is possible to compare the experimentally-detected relation $\gamma'_F = f(n)$ (see Figure 41b, circles) to the relation calculated by Formula (32) (see Figure 41b, solid line). It is clear that the calculated relation is in satisfactory agreement with the experimental relation, thereby explaining its possible deviation from linearity with large n.

It is also interesting to compare our minimum distance R_{min} to the statistical distribution of the ion-ion distances for the nearest neighbors when particles are considered to be point particles. The probability of nearest neighbor detection at a distance R for point particles is given by the expression [170]

$$\varphi(R) = 4\pi R^2 n \exp[-(4/3)\pi R^3 n] \tag{180}$$

and is shown in Figure 43b for the particle concentrations of interest to us. Such a distribution is often used for dilute solutions, finding the mean statistical distance between ions $\bar{T} = 0.9 \, [(4/3)\pi n]^{-1/3}$ from the probability maximum. It is clear that the average distance between ions drops with increasing concentration.

This analysis has demonstrated that the minimum possible distance between Sm^{3+} ions in Li-La-Sm-phosphate glass is very high: R_{min} = 5.6 Å and in a number of cases exceeds the statistical mean $\bar{R}$. In our model of spherical impurity complexes of radius $R_{min}/2$ the forbiddance at $R < R_{min}$ will cause a sharp change in the distribution function and will "nullify" its left (relative to R_{min}) section (represented by the dashed line). We can assume that the ions that would be at distances from 0 through R_{min}: $P_0 = \int_0^{R_{min}}$

$\varphi R)dR$ in the statistical model are at the minimum possible distance R_{min} in our model, thereby forming a spatially (with regard to distance) ordered ion set with fixed distances between neighbors (yet with random angles and numbers of neighbors). The remaining ions comprise a spatially disordered group $P_D = 1 - P_O$, since their distances $R_{ii} > R_{min}$ are permitted in this model and can have random values.

As we see the quantitative relation between these two sets will be determined by the total impurity concentration and R_{min}. It is clear from Figure 44b that if the fraction of ions that will fall within the "forbidden" range according to Expression (180) (R_{min} = 0) (0 < R < 5.6 Å) at a low concentration $n(Sm^{3+})$ = 2.5 • 10^{20} cm^{-3} yet are now at a distance $R = R_{min}$ = 5.6 Å, is relatively small: $P_O = \int_0^{5.6Å} \varphi(R)dR = 0.18$, although as the concentration rises to 9.2 • 10^{20} cm^{-3} $\bar{R}$ it approaches R_{min} and the number of ordered ions (P_O = 0.51) approaches the number of disordered ions (P_D = 0.49).

The ordered set begins to become dominant for n = 2.3 • 10^{21} cm^{-3} since $\bar{R} < R_{min}$ and only a small fraction of the ions remain in the disordered state (P_D = 0.17, P_O = 0.83).

The data given above obtained by integration of the function $\varphi(R)$ are in good agreement with the data provided on the previous page which allows us to use such concepts and to employ the glass models utilized above.

The analysis reveals that at high concentrations the impurity system cannot be described by a chaotic distribution of point particles with the corresponding n or $\bar{R}$; it is more appropriate for an ordered medium with an order parameter R_{min}. The method proposed for determining the microparameter R_{min} permits a specific search for vitreous matrices with large values of R_{min} for using such glasses to fabricate highly concentrated laser media with anomalously weakly-expressed luminescence quenching. Another aspect to this issue is the increasingly significant possibility for investigating the spatial distribution of impurities in various vitreous and liquid media.

Using optical techniques to measure R_{min} and the degree of order in the glass is only the first step in this direction. Advancements will occur more rapidly by using computer processing, and by modeling luminescence decay kinetics when analytic solutions either do not exist or when they are valid, ignoring finite particle dimensions. A comparison of these calculations to experimentally-observed luminescence decay curves may provide new information on the distribution function of active impurities in disordered condensed matter.

7. Donor-Acceptor Interaction of Nd^{3+} Ions in Glass

The most important practical class of structurally- and spectrally-disordered impurity media are laser glasses activated by Nd^{3+} ions. The structural features of the Nd^{3+} electron levels require, in virtually all cases, simultaneous incorporation of both donor-acceptor quenching ion-ion interactions and donor-donor migration interactions. The intermediate $^4I_{13/2}$, $^4I_{15/2}$ energy levels in neodymium between the metastable and ground levels are distributed such that the luminescent and absorption transitions that terminate at these levels may be in resonance which, as in the case of Sm^{3+} ions, makes possible concentration luminescence self-quenching of Nd^{3+} (donor-acceptor transfer) [7, 13, 15, 23]. At the same time the lack of forbidden stimulated electrical dipole $^4F_{3/2}-^4I_{9/2}$ transitions often give rise to a high efficiency of $(Nd^{3+})^*-(Nd^{3+})^*$ migration interactions (donor-donor energy transfer) which commonly exceed the quenching interactions [6, 30, 101, 151, 171].

A rather large number of studies devoted to energy degradation in neodymium-containing media [7, 13, 15, 23] appeared prior to the formulation of our study [85]. However, the most thorough and concise (on the microparameter level) research was carried out on only a few crystalline media with a known lattice structure [7, 15].

Unlike ordered crystalline matrices in which at least one macroparameter R_{min} or the entire structure of the cation sublattice are known it is significantly more difficult to achieve the microlevel in investigating disordered condensed matter.

The most direct method of converting from a phenomenological description of quenching to the elementary interaction microparameters, as demonstrated with the Sm^{3+} ions, is to detect and analyze the static disordered energy transfer stage in the decay kinetics. The analysis of experimental data for Sm^{3+} ions (see Section 6) and computer modeling results (see Section 5.3) show that the equal-probability chaotic ion-ion activator distribution model used in Refs. [82, 83] to obtain analytic expressions (30) and (31) for the kinetics accurately describes this stage through a concentration of $c = 0.20$. The latter, as is clear from Figures 2 and 6, permit independent determination of S and C_{DA} so Expressions (28) for the ordered state can then be used for find R_{min}, thereby obtaining a complete set of macroparameters determining the elementary active ion interaction mechanisms in a disordered medium.

The presence of donor-donor energy migration when acceptors are present substantially enhances quenching of the remote, isolated (from the acceptors) excitations so the final stage of the donor decay becomes exponential. Such dynamical decay ordering over long time periods can substantially narrow or even out the range of observation of static disordered energy transfer which is already limited by remote spatial ordering over short times ($t < t_1$).

One method of expanding the static disordered decay range is to reduce the migration quenching rate by weakening donor-donor interactions $\overline{W}(C_{DD})$. It is possible to use a retarding action such as temperature (cooling to $T = 4.2$ K, see Section 5.1) and an axial magnetic field ($H \approx 39$ kOe [172]) to slow the migration rate among donors. Laser excitation selectivity at a reduced temperature (see Section 5.1) can also substantially reduce migration among the donors. The square-law nature of the concentration dependence for the migration quenching rate ($\overline{W} \sim n^2$) corresponding to the final decay stage (migration-controllable relaxation, Expression (122)) and the linear law for the maximum deactivation rate in the initial, ordered, stage of static quenching ($W_{max} \sim n$ [22]) determine the dynamic range of the decay rate in

the disordered static decay stage: $\overline{W}(n^2) < \left[\gamma_F(n)/t^{\frac{S-3}{S}} \right] < W_{max}(n).$ It is

clear that the best conditions for analyzing the latter are achieved at minimum impurity concentrations ($n \to 0$) at the same time that with n greater than a certain critical level [15, 85]

$$n_{cr} = \frac{\sqrt{C_{DA}}}{\left(\dfrac{2\pi}{3}\right)^{3/2} R_{min}^3 \sqrt{C_{DD}}} \tag{181}$$

the disordered Förster stage will not generally be detected due to convergence of static and dynamic kinetic ordering. At the same time the spontaneous radiative donor excitation decay channel with a finite radiative rate A imposes a lower limit on experimental measurements of the kinetics of nonradiative losses at low concentrations when $W_{max}(n)$; $\gamma_F(n)/t^{(S-3)/S}$; $\overline{W}(n^2)$ is much less than A which is related to the limitations in the experimental accuracy of measurement of the decay kinetics in its processing as well.

These concepts can be illustrated by analyzing the experimental data in Figure 44 which presents the luminescence decay kinetics of Nd^{3+} ions in

Li-La-Nd-phosphate glasses with different activator concentrations measured under laser excitation (γ_E = 567.3 nm) and nonselective recording for T = 4.2 K [85]. The figure reveals that the luminescence decay kinetics are exponential for a specimen of low Nd^{3+} ion concentration (n = 2.7 · 10^{19} cm^{-3}) where the experimental decay rate cannot be differentiated from the spontaneous decay rate, indicating satisfaction of the conditions $\gamma_F(n)/t^{(S-3)/S}$; $\overline{W}(n^2) \ll A$ = 2530 s^{-1}. A near-exponential decay across the entire time scale yet with a substantially greater nonradiative deactivation rate is also observed with increasing Nd^{3+} ion concentration to 1.08 · 10^{21}- 2.7 · 10^{21} cm^{-3}. This indicates satisfaction of the condition $n > n_{cr}$ [15] where the static and dynamic ordering stages converge, forming a single exponential decay law at a rate W_{max}.

The only specimen in which we were able to identify all three nonradiative deactivation stages (or, more important, the static, disordered stage) was glass with an $n(Nd^{3+})$ concentration of 5.4 · 10^{20} cm^{-3}. Analysis of the luminescence decay kinetics of this glass revealed that the kinetics are exponential at t < 400 μs with a quenching rate W_{max} = τ^{-1} - A = 860 s^{-1}. For t > 400 μs the luminescence decay kinetics are accurately described by the law $\exp[-\gamma_F\sqrt{t} - (\overline{W} + A)t]$ (with macroparameters γ_F = 22 s$^{-1/2}$ and $\overline{W}$ = 130 s^{-1}) which, as demonstrated in Refs. [7, 15, 23] is characteristic of dipole-dipole quenching interaction of Nd^{3+} ions. By substituting the macroparameters γ_F, W_{max} and the concentration n = 5.4 · 10^{20} cm^{-3} into Expressions (28) and (31) the techniques from Section 2 can be used to independently determine both elementary quenching interaction microparameters: C_{DA} (Nd-Nd) = 3 · 10^{-2} nm^6/ms, R_{min} (Nd-Nd) = 5.1 Å.

The minimum possible distance of Nd^{3+} ions in Li-La-Nd-phosphate glass corresponds to R_{min} = 5.6 Å which was the value we obtained for an Li-La-Sm-phosphate glass of similar composition and chemical properties (see Section 5.4). Such a high value of R_{min} which is one of the principal causes of anomalously-weak concentration quenching of Nd^{3+} ions, demonstrates the important structural feature of this phosphate glass which, like a number of crystalline phosphates of similar composition, is characterized by an isolated relative configuration of neodymium-oxygen dodecahedrons (NdO_8) [169]: $(R_{min}(NdP_5O_{14})$ = 5.19 Å, R_{min} $(LiNdP_4O_{12})$ = 5.64 Å).

Knowing C_{DA}(Nd-Nd) and R_{min}(Nd-Nd) it is possible to calculate the elementary quenching interaction probability in a pair of Nd^{3+} ions at the minimum possible distance (R_{ij} = R_{min}): W_0 = C_{DA}/R_{min}^6 = 1.7 · 10^3 s^{-1}. It is precisely this quantity that determines the boundary time of the transition from static ordered (exponential) decay to static disordered (Förster) decay, which was equal to t_1 = $(W_0)^{-1}$ = 590 μs (see Figure 44).

Assuming spherical neodymium-oxygen complexes of radius $R_{min}/2$ independent of the Nd^{3+} ion concentration we can estimate the maximum possible neodymium ion concentration in this material from their densest packing conditions with conservation of the anomalously-weak quenching effect and a constant R_{min} (Nd–Nd):

$$n_{max} = \sqrt{2} R_{min}^{-3} \approx 10^{22} \text{ cm}^{-3}.$$

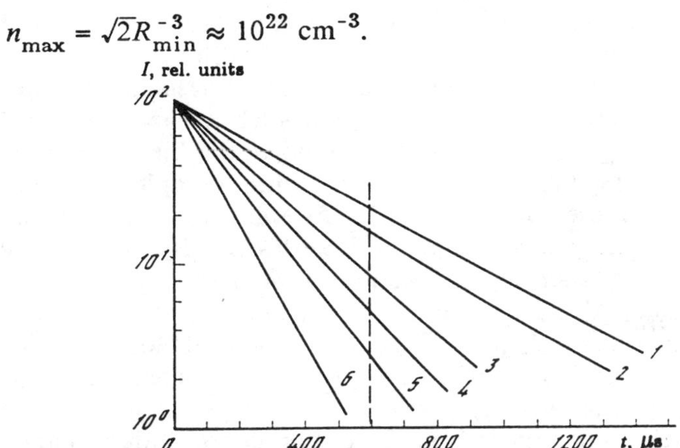

Fig. 44. Kinetics of luminescence decay from the $^4F_{3/2}$ level of Nd^{3+} in Li-La-Nd-phosphate glasses with different Nd^{3+} concentrations at $T = 4.2$ K. n, 10^{20} cm^{-3}: 1 - 0.27, 2 - 5.4, 3 - 11.4-12, 5 - 20.6-27.

The nonresonant nature of donor-acceptor quenching interaction of Nd^{3+} ions with emission of high energy vibrational radiation (800 cm^{-1}) [85] as well as the coincidence of the microparameter $C_{DA} = 3 \cdot 10^{-2}$ nm^6/ms and the quenching macroparameter in crystal phosphates [85] indicate that C_{DA} is relatively insensitive to the composition and inhomogeneous properties of the Nd^{3+} optical centers in this material. This allows us to use Formula (28) to calculate R_{min} (Nd–Nd) based on the values of W_{max} determined from Figure 44 for our specimens by assuming C_{DA} (Nd–Nd) is constant for all glass specimens with different Nd^{3+} concentrations:

$n(Nd^{3+}) \cdot 10^{20}$ cm^{-3}	5.4	8.1	11	12	20	27
W_{max}, s^{-1}	850	1100	1700	2350	3300	5770
R_{min}, Å	5.1	5.4	5.1	4.8	5.0	4.6

The values of R_{min} (Nd–Nd) exceed 5 Å for the majority of glasses with different Nd^{3+} concentrations. The observed drop in R_{min} (Nd–Nd) for glass specimens $n = 1.2 \cdot 10^{21}$ and $2.7 \cdot 10^{21}$ cm^{-3} is most likely due to the fact that the glass specimen may be a nonoptimum specimen due to the duration of its fabrication process aimed at achieving maximum dewatering (to remove the OH$^-$ hydroxyl impurity). The significant dependence of R_{min} (Nd–Nd) on composition is a well-known fact for rare-earth crystalline phsophates [169].

The possibility for measuring (estimating) the migration donor-donor Nd-Nd interaction efficiency without using the selective techniques described above is possible only by introducing an acceptor that is stronger than neodymium in the neodymium-containing medium.

The OH$^-$ hydroxyl ions which are present in laser phsophate glasses and which degrade their properties represent such a quencher, as demonstrated in Refs. [7, 85]. These ions will show up in the blend or glass during the fabrication process due to air exposure, while complex techniques are required to remove these ions. Refs. [7, 85] employed an analysis of the Förster disordered static transfer stage to identify the dipole-dipole character of $Nd^{3+} \rightarrow OH^-$ quenching interactions and to estimate the quenching microparameter $C_{DA}(Nd-OH) \approx 0.6$ nm^6/ms which was more than an order of magnitude higher than $C_{DA}(Nd-Nd)$ and turned out to be temperature-dependent in the 4.2-300 K temperature range and insensitive to the inhomogeneous nature of the Nd^{3+} spectra.

Estimates and a comparison of the donor-donor (Nd-Nd) and donor-acceptor (Nd-OH) interaction microparameters demonstrates that the hop energy migration conditions hold at $T \geqslant 77$ K. However, a lack of information on the microparameter R_{min} (Nd-OH) which may differ substantially from R_{min} (Nd-Nd), makes it impossible to calculate n_{cr} in advance; this parameter separates the migration-controllable quenching range where $\overline{W} \sim n_D n_A$ from the fast migration range with saturation of donor-acceptor transfer for which $W_{max} \sim n_A$.

At the same time an experimental investigation of the dependence of the energy migration rate on the product of the donor and acceptor concentrations $\overline{W} = f[n(Nd)n(OH)]$ has demonstrated [85] (see Figure 45) a linear relation up through donor concentrations $n_D(Nd) = 2.7 \cdot 10^{21}$ cm^{-3} which represents a criterion for the absence of supermigration (the kinetic limit) which is manifested as a saturation of the dependence of W on $n_D(Nd)$ [22, 28]. This makes it possible to write a lower estimate for the critical concent-

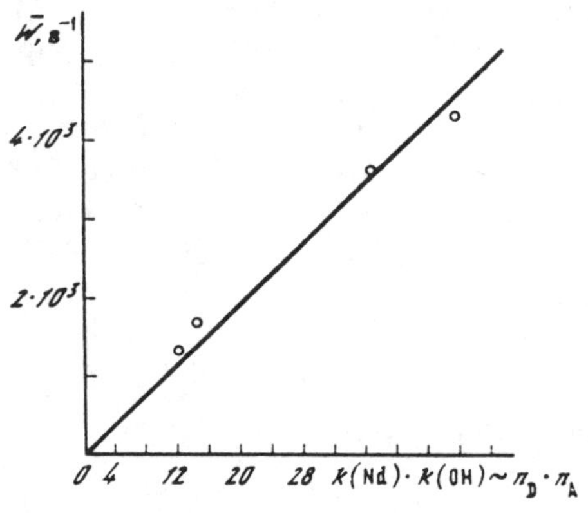

Fig. 45. Migration quenching probability plotted as a function of the product of the n_D donor concentration $n_D(Nd^{3+}) \sim k(Nd^{3+})$ and the acceptor concentration: $n_A(OH) \sim k(OH)$: $\overline{W}(Nd \rightarrow Nd \rightarrow OH) = f(k(Nd)k(OH))$.

ration for the Nd $\rightarrow$ Nd $\rightarrow$ OH quenching channel: $n_{cr} > 2.7 \cdot 10^{21}$ cm^{-3}.

The hop mechanism of migration quenching of Nd^{3+} ion excitations by OH$^-$ vibrations established here presents us with the interesting possibility for determining the average donor-donor interaction microparameter C_{DD} without precise knowledge of n_A and C_{DA}. Indeed, the uniform nature of the functional relation between the quenching macroparameters in the Förster disordered stage (γ_{DA}) and the migration-controllable hop migration stage ($\overline{W}$) with the microparameter C_{DA} and acceptor concentration n_A

$$\gamma_{DA} = {}^4/_3\, \pi^{3/2} n_A \sqrt{C_{DA}}, \tag{182}$$

$$\overline{W} = {}^4/_9\, \pi^{7/2} n_D \sqrt{C_{DD}}\; n_A \sqrt{C_{DA}} \tag{183}$$

makes it possible to expression one experimentally-measured quantity ($\overline{W}$) through another (γ_{DA}) eliminating the explicit form of the dependence of $\overline{W}$ on n_A and C_{DA}:

$$\overline{W} = \frac{\pi^2}{3}\, n_D \sqrt{C_{DD}}\; \gamma_{DA}. \tag{184}$$

This is an important formula for application to experiments where the acceptor is an uncontrolled impurity whose concentration n_A and, therefore, the quenching efficiency C_{DA} are not known exactly and can be only roughly approximated and even not in all cases.

Substituting the neodymium concentration $n_D = 1.1 \cdot 10^{21}$ cm^{-3} into the last expressions and the measured quenching macroparameter values $\gamma_{DA} = 6$ s$^{-1/2}$ and $\overline{W} = 1700$ s^{-1} we find the donor-donor interaction efficiency macroparameter without even knowing the precise hydroxyl ion concentration in the glass. C_{DD} was equal to 6 nm^6/ms which roughly corresponds to the values obtained by selective spectroscopy [6, 146, 151]. It is important to note that this quantity is only a certain effective average quantity since it represents an estimate using Formula (183) for the case of the homogeneously-broadened spectra of the interacting donors and it does not account for the spectral inhomogeneities or the possible frequency dependence of Nd-Nd interaction efficiency in glass at $T = 77$ K (see Section 5.2).

Expression (181) clearly shows that the value of C_{DD}, C_{DA}, R_{min} [85] allows us to use the lower estimate given above for the critical concentration for Nd-Nd-Oh quenching, $n_{cr} > 2.7 \cdot 10^{21}$ cm^{-3} to find the upper estimate for the minimum distance of Nd^{3+} ions and OH$^-$ hydroxyl ions in the Li-La-Nd-phosphate glass under test. It turned out that R_{min} (Nd-OH) < 3.1 Å. The significantly lower value of this quantity compared to R_{min} (Nd-Nd) $= 5.1$ Å is understandable since the ion of the OH$^-$ hydroxyl group is not itself an isolated complex like, for example, NdO$_8$, and can close on the latter, interacting at a distance of the order of R (Nd-O) $= 2.0$-2.5 Å.

Investigating the luminescence decay via the Nd$-$Nd-OH channel in specimens of concentration $n > 2.7 \cdot 10^{21}$ cm^{-3} (which we expect is possible since $n_{max} = \sqrt{2}/R_{min}^3 = 10^{22}$ cm^{-3}) will make it possible to refine R_{min} (Nd-OH) by quenching measurements under kinetic limit conditions (fast migration). Exact knowledge of this parameter may prove useful in crystal chemistry in understanding the structure of glass and the aspects of introducing various impurities in glass.

Conclusion

This paper has focused on a wide variety of noncoherent quenching and migration nonradiative energy transport processes both in the case of a pair of interacting particles and under conditions of collective relaxation in an ensemble of spatially- and spectrally-disordered particles. These processes play an important role in such fundamental phenomena of the optics of condensed matter as concentration quenching and luminescence sensitization, anti-Stokes luminescence, spectral migration, and delocalization of optical excitations. At the same time they determine the important properties of activated crystals and gases when used as laser and luminescing media in quantum electronics.

We employed luminescence studies of Yb^{3+}, Nd^{3+}, Eu^{3+}, and Sm^{3+} rare-earth ions in laser and other model glasses to demonstrate experimental techniques and the capabilities of nonstationary kinetic spectroscopy for investigating excitation quenching and migration which have reached not only the qualitative but also the quantitative level today.

Time-resolved luminescence and excitation spectra together with precision measurement of the luminescence decay kinetics under selective excitation and recording conditions have made it possible to identify both the spectral and kinetic features of direct and indirect phonon-assisted energy transfer and energy transfer associated with certain aspects of the spatial activator distribution in solids. Cases where interactions of higher multipolarity (dipole-quadrupole for Sm^{3+} ions and quadrupole-quadrupole for Eu^{3+} ions) predominate were established for the first time based on an analysis of the nonstationary energy transfer stage. The identification and investigation of the initial ordered transfer kinetic stage made it possible to determine for the first time the energy parameter — the minimum distance of active ions — for disordered vitreous solids; this turned out to be highly dependent on glass composition. The significant dynamic measurement range permitted detection of a diffusion dipole-dipole excitation delocalization stage for a disordered ensemble of particles as well as the concentration and spectral relations of the rates in this stage. We emphasize that the set of precision kinetic luminescent spectroscopy data for the case of selective laser excitation combined with a comprehensive analysis of such data based on modern theoretical models and computer statistical modeling results make it possible to not only identify the common qualitative mechanisms of these transfer processes but also to identify the most important quantitative characteristics which determine the character, efficiency, and probability of interactions on the microlevel.

References

1. V. M. Agranovich and M. D. Galanin, *Electron Excitation Energy Transfer in Condensed Matter. Modern Problems in Condensed Matter Sciences*, vol. 3, Amsterdam, North-Holland.

2. V. L. Ermolaev, E. N. Budonov, E. B. Sveshnikova, et al., *Nonradiative Transfer of Electron Excitation*, Leningrad: Nauka, 1977, 311 pp. (in Russian).

3. A. A. Kaminskii, V. V. Osiko, "Inorganic laser Materials with a Thin Structure", *Bull. Acad. Sci. USSR, Ser. Inorgan. Mat.*, vol. 1, 1965, pp. 2049-2087; vol. 3, 1967, pp. 417-463; vol. 6, 1970, pp. 629-696.

4. A. A. Kaminskii, *Laser Crystals*, Moscow: Nauka, 1975, 225 pp. (in Russian).

5. G. O. Karapetyan and A. L. Reishahrit, "Luminescent Glasses as a Laser Medium", *Bull. Acad. Sc. USSR, Ser. Inorgan. Mat.*, vol. 3, 1967, pp. 217-259.

6. N. E. Alekseev, V. P. Gapontsev, M. E. Zhabotinskii, et al., *Laser Phosphate Glasses*, Moscow: Nauka, 1980, 352 pp. (in Russian).

7. I. A. Shcherbakov, *Investigation of Excitation Energy Relaxation Processes in Crystals and Glasses Doped with Rare-Earth Ions"*, PhD Dissertation in Physics and Math., FIAN SSSR, 1978, 358 pp. (in Russian).

8. M. J. Weber, "Laser Excited Fluorescence Spectroscopy in Glass", *Laser Spectroscopy of Solids*, Ed. by W. W. Yen and P. M. Selzer, Berlin: Springer, 1981, pp. 189-239.

9. L. G. van Vitert and L. F. Johnson, "Energy Transfer Between Rare-Earth Ions", *J. Chem. Phys.*, vol. 44, 1966, pp. 3514-3521.

10. Ivey, H. F., "Sensitized Luminescence and its Application in Laser Materials, *Proc. Intern. Conf. Luminescence*, Budapest, 1966, pp. 2027-2049.

11. M. N. Tolstoi, "Nonradiative Energy Transfer Between Rare-Earth Ions in Crystals and Glasses", In: *Spectroscopy and Crystals*, Ed. by S. V. Grum-Grzhimailo, B. S. Skorobogatov, P. P. Feofilov, and V. I. Cherepanov, Moscow: Nauka, 1970, pp. 124-135 (in Russian).

12. *Paper of the First All-Union Seminar on Energy Transfer in Condensed Matter*, Yerevan: Acad. Sci. ArmSSR, 1970, 196 pp.

13. T. T. Basiev, Yu. K. Voron'ko, T. G. Mamedov, et al., "Excitation Relaxation Processes of Metastable Levels of Rare-Earth Ions in Crystals", In: *Spectroscopy of Crystals*, Ed. by A. A. Kaminskii, Z. A. Morgenshtern, and D. T. Sviridov, Moscow: Nauka, 1975, pp. 155-184 (in Russian).

14. G. M. Zverev, I. I. Kuratev, and A. M. Onishchenko, "Nonradiative Excitation Energy Transfer Between Trivalent Rare-Earth Ions in Crystals", *Ibid.*, pp. 184-192 (in Russian).

15. Yu. K. Voron'ko, T. G. Mamedov, V. V. Osiko, et al., "Nature of Nonradiative Excitation-Energy Relaxation in Condensed Media at High Activator Concentrations" *Sov. Phys. JETP*, vol. 44, 1976, p. 251.

16. T. T. Basiev, Yu. K. Voron'ko, V. V Osiko, et al., "Experimental Observation of 'Excitation Trapping' in a System of Strongly Interacting Particles", *Ibid.*, vol. 43, p. 637.

17. I. A. Bondar, A. I. Burshtein, A. V. Krutikov, et al., "Investigation of the Processes of Electron Excitation Relaxation in Crystals for Arbitrary Relations Between Interaction Microparameters and Concentrations of Donors and Acceptors", *Ibid.*, vol. 54, 1981, p. 45.

18. W. W. Yen, P. M. Selzer, "High Resolution Laser Spectroscopy of Ions in Crystals", *Laser Spectroscopy of Solids*, Ed. by W. W. Yen and P. M. Selzer, Berlin: Springer, 1981, pp. 141-188.

19. S. I. Golubov and Yu. V. Konobeev, "The Averaging Technique in the Theory of Resonance Electron Excitation Energy Transfer", *Sov. Phys. Solid State*, vol. 13, 1971, p. 2679.

20. V. P. Sakun, "Kinetics of Energy Transport in Crystals", *Ibid.*, vol. 14, 1972, p. 1906.

21. A. S. Agabekyan, "Theory of Energy Resonant Transfer Accounting for Inhomogeneous Spectral Line Broadening", *Opt. And Spectrosc.*, vol. 30, 1971, pp. 449-455.

22. A. I. Burshtein, "Hopping Mechanism of Energy Transfer", *Sov. Phys. JETP*, vol. 35, 1972, p. 882.

23. M. V. Artamonova, I. M. Briskina, A. I. Burshtein, et al., "Investigation of the Time Dependence of Ion Luminescence and Estimation of Electron Excitation Migration Among These Ions in Glass", *Ibid.*, pp. 863-871.

24. L. D. Zusman, A. I. Burshtein, "Kinetics of Intermolecular Energy Transfer in Condensed Matter", *Zh Prik. Spek.*, vol. 15, 1971, pp. 124-129.

25. L. D. Zusman, "Luminescence Quenching from Excitation Migration in Solid Solutions", *Opt. and Spectrosc.*, vol. 36, 1974, pp. 497-502.

26. M. M. Rikenglaz and I. M. Rozman, "The Luminescence Kinetics of Rigid Solutions with Energy Transfer", *Ibid.*, pp. 100-103.

27. A. I. Burshtein, "Quasiresonant Energy Transfer. Part 1. Statistical

Quenching of Luminescence" *Avtometriya*, no. 5, 1978, p. 65-84, no. 6, p. 72-90 (in Russian).

28. D. L. Huber, "Dynamics of Incoherent Transfer", *Laser Spectroscopy of Solids"*, Ed. by W. W. Yen and P. M. Selzer, Berlin: Springer, 1981, pp. 83-811.

29. R. Beigang, G. Litfin, and H. Welling, "Frequency Behavior and Line-Width of CW Single Mode Color Center Lasers", *Opt. Communs.*, vol. 22, 1977, pp. 269-271.

30. A. K. Przhevuskii, "Inhomogeneous Spectral Structure of Glasses Activated by Rare-Earth Ions", In: *Spectroscopy of Crystals*, Ed. by P. P. Feofilov, Leningrad: Nauka, 1978, pp. 96-108 (in Russian).

31. T. T. Basiev, Yu. K. Voron'ko, V. V. Osiko, and A. M. Prokhorov, "Laser Spectroscopy of Doped Crystals and Glasses", In: *Spectroscopy of Crystals*, Ed. by A. A. Kaplyanski, Leningrad, Nauka, 1983, pp. 57-82 (in Russian).

32. A. K. Przhevuskii, "Invariant Parameters and Statistical Modeling of the REE Optical Centers in Glasses", *Ibid*, pp. 82-95 (in Russian).

33. Th. von Förster, Zwischenmolekulare Energiewanderund und Fluoreszenz, *Ann. Phys.*, vol. 2, 1948, pp. 53-75 (in German).

34. Th. von Förster, Experimentelle und theoretische Untersuchung des zwischenmolekularen Übergangs von Elektronenanregungsenergie, *Ztschr. Naturforsch. A*, vol. 4, 1949, p. 321-327 (in German).

35. D. L. Dexter, A Theory of Sensitized Luminescence in Solids, *J. Chem. Phys.*, vol. 21, 1953, pp. 836-850.

36. A. S. Davydov, *Quantum Mechanics*, 1963, Moscow: Fizmatgiz, 350 pp. (in Russian).

37. A. S. Davydov, "The Radiationless Transfer of Energy of Electronic Excitation Between Impurity Molecules in Crystals", *Phys. Status Solidi*, vol. 30, 1968, pp. 357-366.

38. A. S. Davydov, "Electron Excitations and Lattice Vibrations in Molecular Crystals" *Bull. Acad. Sci. USSR, Phys. Ser.*, vol. 34, 1970, p. 483.

39. D. L. Dexter, Th. Förster, and R. S. Knox, "The Radiationless Transfer of Energy of Electron Excitation Between Impurity Molecules in Crystals", *Phys. Status Solid i*, vol. 34, 1969, p. K159-K162.

40. A. I. Burshtein, "Kinetics of the Transfer Processes in a System Consisting of Three Levels", *Teoret. i Eksperim, Khimiya*, vol. 1, 1965, p. 563 (in Russian).

41. V. Ya. Gamurar, Yu. E. Perlin, and B. S. Tsukerblat, "Nonradiative Energy Transfer in Activated Crystals", *Sov. Phys. Solid State*, vol. 11, 1969, Rus. p. 1193.

42. Sh. Kudzhmauskas and I. V. Batarunas, "Theory of Resonance Interaction Between Identical Impurity Centers Through the Phonon Field Taking Account of Energy Dissipation", *Litov. Fiz. Zh.*, vol. 9, 1969, p.

721 (in Russian).

43. E. D. Trifonov and V. L. Shekhtman, "On the Theory of Nonradiative Transitions", *Sov. Phys. Solid State*, vol. 11, 1969, Rus. p. 2984.

44. A. S. Agabekyan and A. O. Melikyan, "Investigation of Nonradiative Energy Transfer by the Density Matrix Technique", *Paper of the All-Union Seminar on Nonradiative Energy Transfer in Condensed Matter* (Lory 1969, Yerevan, Acad. Sci. Arm. SSSR) 1970, p. 5 (in Russian).

45. E. F. Kustov and L. I. Surogin, "Energy Excitation Transfer Process Taking Account of Transverse and Longitudinal Relaxation", *Ibid.*, p. 17 (in Russian).

46. E. D. Trifonov, "Theory of Excitation Transfer in Crystals with Impurity Centers", *Bull. Acad. Sci. USSR, Ser. Phys.*, vol. 35, 1975, Rus. p. 1330.

47. A. S. Agabekyan and A. O. Melikyan, "On the Applicability Criteria of the Theory of Resonance Energy Transfer", *Opt. and Spectrosc.*, vol. 32, 1972, Rus. p. 288.

48. E. F. Kustov, "Excitation Transfer Between Electron Levels of Impurity Centers in Crystals", *Ibid.*, Rus. p. 323.

49. V. L. Shechtman, E. D. Trifonov, *Phys. Status Solidi*, vol. B41, 1970, pp. 855-860.

50. V. L. Shekhtman, "Investigation of the Theory of Nonradiative Energy Transfer of Electron Excitation in Doped Crystals", *Abstract of PhD Dissertation, Phys.-Math*, Leningrad, 1973, 24 pp. (in Russian).

51. A. P. Bozhko and K. K. Pukhov, "Energy Transfer in Condensed Matter", *Sov. Phys. Solid State*, vol. 15, 1973, Rus. p. 980.

52. E. D. Trifonov and V. L. Shekhtman, "On the Theory of Energy Transfer in Doped Crystals", In: *Spectroscopy of Crystals*, Ed. by P. P. Feofilov, Leningrad: Nauka, 1973, p. 193 (in Russian).

53. A. S. Davydov and A. A. Serikov, "Theory of Electron Excitation Energy Migration in Crystals", *Bull. Acad. Sci. USSR, Phys. Ser.*, vol. 37, 1973, Rus. p. 474.

54. E. D. Trifonov and V. L. Shekhtman, "The Theory of Nonradiative Transitions in Crystals with Impurities", In: *Physics of the Impurity Centers in Crystals*, Ed. by G. S. Savt, Tallin: Acad. Sci. Estimate. SSR, 1972, p. 350.

55. Sh. P. Kudzhmauskas and I. V. Batarunas, "Nonradiative Electron Excitation Energy Transfer in Doped Crystals", *Ibid.*, p. 585 (in Russian).

56. A. I. Burshtein and A. Yu. Pusep, "Collective Quenching of Luminescence", *Sov. Phys.-Solid State*, 1974, vol. 16, Rus. p. 2318.

57. Yu. M. Homenko, "The Theory of Cross-Relaxation in Donor-Acceptor Systems", *Ukrain. Fiz. Zh.*, vol. 20, 1957, p. 1440 (in Russian).

58. D. L. Dexter and J. H. Schulman, "Theory of Concentration Quenching in Inorganic Phosphors", *J. Chem. Phys.*, vol. 22, 1954, pp. 1063-1070.

59. T. Holstein, S. K. Lyo, and R. Orbach, "Excitation Transfer in Disordered Systems", In: *Laser Spectroscopy of Solids*, Ed. by W. M. Yen and P. M. Selzer, Berlin, etc.: Springer, 1981, pp. 39-82.

60. Personov, R. I., "Selective Spectroscopy of Complex Molecules in Solutions and its Applications", In: *Spectroscopy and Excitation Dynamics of Condensed Molecular Systems*, Ed. by V. M. Agranovich and R. M. Hochstrasser, *Modern Problems in Condensed Matter Sciences, Vol. 4*, Amsterdam: North Holland, 1983, Ch. 4.

61. A. A. Maradudin, *Theoretical and Experimental Aspects of the Effects of Point Defects and Disorder on the Vibrations of Crystals*, New York, London: Acad. Press Inc., 1966.

62. K. K. Rebane, *Elementary Theory of Vibrational Spectral Structure of Impurity Centers in Crystals*, Moscow: Nauka, 1968, 232 pp. (in Russian).

63. I. S. Osad'ko, "Investigation of Vibronic Interaction Based on the Structural Optical Spectra of Impurity Centers", *Sov. Phys. Uspekhi*, vol. 128, 1979, p. 31 (in Russian).

64. I. V. Ignatyev and V. V. Ovsyankin, "Vibronic Spectra and Dynamics of Crystals with Rare-Earth Ions", In: *Spectroscopy of Crystals*, Ed. by A. A. Kaplyanskii, Leningrad: Nauka, 1983, p. 36 (in Russian).

65. Yu. E. Perlin and B. S. Tsukerblutt, *Effects of Vibronic Interaction on the Optical Spectra of Paramagnetic Impurity Ions*, Kishinev: Shtiintsa, 1974, 368 pp. (in Russian).

66. I. V. Vasil'ev, G. M. Zverev, G. Ya. Kolodnyi, and A. M. Onishchenko, "Nonresonance Excitation Energy Transfer Between Impurity Rare-Earth Ions", *Sov. Phys. JETP*, vol. 29, 1969, p. 69.

67. R. Orbach, "Phonon Sidebands and Energy Transfer", In: *Optical Properties of Ions in Crystals*, Ed. by H. M. Crosswide and H. W. Moos, New York: Intersci. Pub., 1967, pp. 445-455.

68. V. I. Aleksandrov, A. P. Bozhko, and K. K. Puhov, "Phonon-Assisted Electron Excitation Energy Transfer", *The Second Seminar on Spectroscopic Properties of Luminophors*, 1969, Moscow, p. 34 (in Russian).

69. T. Miyakawa and D. L. Dexter, "Phonon Sidebands, Multiphonon Relaxation of Excited States and Phonon-Assisted Energy Transfer Between Ions in Solids", *Phys. Rev. B. Solid State*, vol. 1, 1970, pp. 2961-2969.

70. A. P. Bozhko and K. K. Pukhov, "Energy Transfer in Condensed Matter", *Sov. Phys. Solid State*, vol. 15, 1973, Rus. p. 980.

71. E. N. Budunov and V. L. Shekhtman, "The Theory of Energy Transfer in Doped Crystals, *Sov. Phys. Solid State.*, vol. 12, 1970, Rus. p. 2809.

72. Yu. E. Perlin, S. I. Boldyrev, and V. N. Enakii, "Many-Phonon Excitation Transfer and Intracenter Relaxation Processes in Doped Crystals", *Bull. Acad. Sci. USSR, Phys. Ser.*, vol. 43, 1979, p. 170.

73. R. Silsbee, "Thermal Broadening of the Mossbayer Lines and Narrow Line Electron Spectra in Solids", *Phys. Rev.*, vol. 128, 1962, pp. 1726-1732.

74. M. A. Krivoglaz, "Theory of Phonon-Free Line-Broadening in Mössbauer or Optical Spectra", *Sov. Phys. Solid State*, vol. 6, 1964, Rus. p. 1707.

75. D. E. McCumber, "Width of the Narrow No-Phonon Lines in the Optical Spectra of Impurities in Solids", *Phys. Rev.*, vol. 128, 1962, pp. 1726-1732.

76. I. S. Osad'ko, "Theory of Temperature Broadening and Shift of Phonon-Free Optical Lines of an Impurity with Arbitrary Electron-Phonon Interaction Values", *Sov. Phys. Solid State*, vol. 14, 1972, Rus. p. 2927.

77. I. S. Osad'ko, "Zero-Phonon Lines and Phonon Wings in Absorption and Fluorescence Spectra of an Impurity", *Ibid.*, vol. 17, 1975, Rus. p. 3180.

78. S. K. Lyo and R. Orbach, "Homogeneous Fluorescence Line-Width for Amorphous Hosts", *Phys. Rev. B. Solid State*, vol. 22, 1980, pp. 4223-4225.

79. I. S. Osad'ko, "Anomalous Thermal Broadening of the Optical Lines of Impurity Centers in a Glass", *Sov. JETP Lett.*, vol. 33, 1981, p. 626.

80. T. L. Reineche, "Fluorescence Line-Width in Glasses", *Solid State Communs.*, vol. 32, 1979, pp. 1103-1106.

81. M. D. Galanin, "Effect of Concentration on the Luminescence of Solutions", *Sov. Phys. JETP*, vol. 28, 1955, p. 485 (in Russian).

82. B. Ya. Sveshnikov and V. I. Shirokov, "Change in the Mean Lifetime and Luminescence Efficiency in the Quenching Process as a Function of the Law of Molecular Interaction", *Opt. and Spectrosc.*, vol. 12, 1962, Rus. p. 576.

83. M. Ynokuti and F. Hirayama, "Influence of Energy Transfer by the Exchange Mechanism on Donor Luminescence", *J. Chem. Phys.*, vol. 43, 1965, pp. 1978-1989.

84. T. T. Basiev, Yu. K. Voron'ko, and A. M. Prokhorov, "Structural Investigation of the Inhomogeneously-Broadened Spectra of RE Ions and Energy Migration Over an Inhomogeneous Contour by Selective Laser Excitation", In: *Spectroscopy of Crystals*, Ed. by P. P. Feofilov, Leningrad, Nauka, 1978, p. 83 (in Russian).

85. A. G. Avanesov, T. T. Basiev, Yu. K. Voron'ko, et al., "Investigation of Electron Excitation Deactivation and Energy Transfer Processes of Neodymium in Higher Concentration Phosphate Glasses", *Sov. Phys. JETP*, vol. 50, 1979, p. 886.

86. O. K. Alimov, M. Kh. Ashurov, T. T. Basiev, et al., *Kinetic Luminescent Spectroscopy of RE Ions in Crystals and Glasses by Selective Laser*

197

Excitation, Preprint of the Lebedev Physics Institute, No. 172, 1980, Moscow, 24 pp. (in Russian).

87. A. G. Avanesov, T. T. Basiev, Yu. K. Voron'ko, et al. "Investigation of the Spatial Distribution of Impurities in Solids by Kinetic Luminescent Spectroscopy, *Sov. Phys. JETP*, vol. 57, 1983, p. 596.

88. T. T. Basiev, Yu. K. Voronko, V. V. Osiko, et al., "Laser Spectroscopy of Crystals with Impurity Ions and Color Center", *Intern. Conf. Defects Insulat. Cryst.: Abstr.*, Riga: Zinatne, 1981, pp. 354-355.

89. A. I. Burstein, "Energy Transfer Kinetics in Disordered Systems", *J. Luminescence.*, vol. 34, 1985, pp. 167-168.

90. A. I. Burshtein, "The Concentration of Quenching of Noncoherent Excitations in Solutions", *Sov. Phys. Usp.*, vol. 27, 1984, p. 579.

91. E. N. Bodunov, "Incorporation of Molecular Sizes in the Theory of Energy Transfer in Solutions", In: *Paper of the Second Repub. Conf. of Junior Scientists on Physics*, Minsk, 1974, no. 2, p. 3 (in Russian).

92. M. D. Galanin, "Excitation Energy Transfer in Luminescent Solutions", In: *Proceedings of the Lebedev Physics Institute*, vol. 12, 1960, p. 3 (in Russian).

93. T. T. Basiev, Yu. K. Voron'ko, and I. A. Shcherbakov, "Effect of Spectral Line Broadening on Electron Excitation Migration Among Impurity Centers in Crystals", *Sov. Phys. JETP*, vol. 39, 1974, p. 1042.

94. T. T. Basuiev, Yu. K. Voron'ko, T. G. Mamedov, and I. A. Shchebakov, "Energy Migration Among Yb^{3+} Ions in Garnet Crystals", *Sov. J. Quant. Electron.*, vol. 5, 1975, p. 1182.

95. D. L. Huber, D. S. Hamilton, and B. Barnet, "Time-Dependent Effects in Fluorescent Line Narrowing", *Phys. Rev. B. Solid State*, vol. 16, 1977, pp. 4642-4650.

96. A. G. Kofman and A. I. Burshtein, "Effect of Spectral Migration on the Shape of an Inhomogeneously-Broadened Luminescence Line", *Sov. Phys. Solid State*, vol. 15, 1973, Rus. p. 2114.

97. A. G. Cabesas and R. P. Treat, "Effects of Spectral Hole Burning and Cross-Relaxation on the Gain Saturation of Laser Amplifiers", *J. Appl. Phys.*, vol. 37, 1966, pp. 3556-3563.

98. V. R. Belan, I. M. Briskina, V. V. Grigoriants, and M. L. Gurari, "Investigation of Laser Luminescence Spectral Line Broadening", In: *Inhomogeneous Spectral Line Broadening of the Laser Active Media*, Kiev: IFAN Ukr. SSR, 1969, p. 151 (in Russian).

99. V. R. Belan, I. M. Briskina, V. V. Grigoriants, and M. E. Zhabotinskii, "Energy Transfer Between Neodymium Ions in Glass", *Sov. Phys. JETP*, vol. 30, 1970, p. 627.

100. V. V. Grigoryants, "Information Content of Luminescence of the Active Medium of a Free-Running Laser", *Sov. Phys. JETP*, vol. 31, 1970, p. 853.

101. V. V. Grigoryants, *Luminescence of Laser Media Under Coherent Irradiation*, Dissertation for Doctorate Degree in Phys. Math., Moscow, 1971 (in Russian).

102. V. V. Grigoryants, M. E. Zhabotinski, and V. B. Markuchev, "Lasing and Spectral-Line Characteristics in Phosphate Glasses and Inorganic Liquid with Neodymium", *IEEE J. Quant. Electron,* vol. 8, 1972, p. 196-198.

103. L. S. Kornienko, P. P. Pashinin, and A. M. Prokhorov, "The Spin-Lattice Relaxation Time of Ti^{3+} in Corundum", *Sov. Phys. JETP,* vol. 15, 1962, p. 45.

104. D. N. Daraseliya and A. A. Manenko, The "Freezing" of Cross-Relaxation in Inhomogeneously-Broadened EPR Lines", *JETP Letters,* vol. 11, 1970, p. 337 (in Russian).

105. A. S. Epifanov and A. A. Manenkov, "Theory of Relaxation in Inhomogeneously-Broadened EPR Lines", *Sov. Phys. JETP,* vol. 33, 1971, p. 976.

106. N. Montegy and S. Shinoya, "Excitation Migration Among Inhomogeneous Broadened Levels of Eu^{3+} Ions", *J. Luminescence,* vol. 8, 1973, pp. 1-17.

107. T. T. Basiev, *Electron Excitation Transfer Between Rare-Earth Ions in Laser Matrices*, Dissertation for Doctorate Degree in Phys. Math., Moscow: FIAN, 1976 (in Russian).

108. O. K. Alimov, T. T. Basiev, Yu. K. Voron'ko, et al., "Laser Spectroscopy of Inhomogeneously-Broadened Lines in Glasses and Associated Electron Excitation Migration", *Sov. Phys. JETP,* vol. 45, 1977, p. 690.

109. T. T. Basiev, Yu. K. Voron'ko, A. Ya. Karasik, et al., "Electron Excitation Spectral Migration Among Nd^{3+} Ions in CaF_2-YF_3 Crystals Under Selective Laser Excitation Conditions", *Sov. Phys. JETP,* vol. 48, 1978, p. 32.

110. W. M. Yen, S. S. Sussman, J. A. Paisner, and M. J. Weber, *Fluorescence Line Narrowing and Energy Migration in up Converted Transition Eu^{3+} Glass*, Lawrence Livermore Lab. Prepr. UCRL-76481, 1975, 16 pp.

111. D. L. Huber, "Fluorescence in the Presence of Traps", *Phys. Rev. B Solid State,* vol. 20, 1979, pp. 2307-2314.

112. W. Y. Ching, D. L. Huber, and B. Barnet, "Models for the Time Development of Spectral Transfer in Disordered Systems", *Ibid.,* vol. 17, 1978, pp. 5025-5028.

113. B. E. Vugmeister, "Spin Diffusion in Disordered Paramagnetic Systems", *Sov. Phys. Solid State,* vol. 18, 1976, Rus. p. 819.

114. S. W. Haan and H. Zwanzig, "Förster Migration of Electronic Excitation Between Randomly Distributed Molecules", *J. Chem. Phys.,* vol. 68, 1978, pp. 1879-1883.

115. C. R. Gouchanour, H. C. Andersen, and N. D. Fayer, "Electronic Ex-

cited State Transport in Solution", *Ibid.*, 1979, vol. 70, pp. 4254-4271.

116. K. Godzik and J. Jortner, "Dispersive Diffusion of Electronic Energy in an Impurity Band", *Chem. Phys. Lett.*, vol. 63, 1979, pp. 428-432.

117. K. Godzik and J. Jortner, "Electronic Energy Transport in Substitutionally Disordered Molecular Crystals", *J. Chem. Phys.*, vol. 72, 1980, pp. 4471-4486.

118. A. Blumen, J. Klafter, and R. Silbey, "Theoretical Studies of Energy Transfer in Disordered Condensed Media", *Ibid.*, pp. 5320-5332.

119. J. Klafter and R. Silbey, "On Electronic Energy Transfer in Disordered Systems", *Ibid.*, pp. 843-848.

120. J. Bernasconi, S. Alexander, and R. Orbach, "Classical Diffusion in One-dimensional Disordered Lattice", *Phys. Rev. Lett.*, 1978, vol. 41, pp. 185-187.

121. J. Heinrichs and N. Kumar, "Spectral Diffusion in Random Lattice", *Phys. Rev. B. Solid State*, vol. 20, 1979, pp. 1377-1389.

122. F. S. Dzheparov, "Semiphenomenological Equations for the Description of Cross-Relaxation in the Disordered System ^{8}Li-^{6}Li in LiF", In: *Radiospectroscopy, Proc. All-Union Symp. on Magnetic Resonance* Ed. by I. G. Shaposhnikov, Perm: Perm State University, 1979, p. 135 (in Russian).

123. F. S. Dzheparov, V. S. Smelov, and V. E. Shestopal, "Concentration Expansion in the Theory of Excitation Migration Along a Disordered Lattice", *Sov. Phys. Lett.*, vol. 32, 1980, p. 47.

124. F. S. Dzheparov, "Migration of Spin Polarization in Certain Disordered Systems", In: *Radiospectroscopy, Interinst. Collect. Sci. Papers*, Perm, p. 20 (in Russian).

125. O. K. Alimov, M. Kh. Ashurov, T. T. Basiev, et al., *Spectrally-Nonselective Energy Migration in Disordered Media*, Preprint of the Lebedev Physics Institute No. 160, 1983, (in Russian).

126. M. D. Galanin and Z. A. Chizhikova, "Energy Excitation Transfer in Crystals of Antracene Doped by Naphthacene", *Opt. and Spectrosc.*, vol. 1, 1956, Rus. p. 175.

127. J. Heber, "Excitation-Diffusion-Limited Energy Transfer in Solids", *Phys. Status Solide. B*, vol. 48, 1971, pp. 319-346.

128. V. L. Shekhtman, "Effect of Exciton Diffusion on Their Energy Transfer to Impurity Centers in Crystals". Part I, *Opt. and Spectrosc.*, vol. 33, 1972, Rus. p. 284.

129. V. M. Agranovich, *Theory of Excitons*, Moscow: Nauka, 1968, 200 pp. (in Russian).

130. V. M. Agranovich, E. P. Ivanova, and S. Sh. Nikolaishvili, "Molecular Exciton Trapping by Impurities and Singlet-Triplet Exciton Annihilation", *Sov. Phys. Solid State*, vol. 15, 1973, Rus. p. 2701.

131. M. Yokota, O. Tanimoto, "Effects of Diffusion on Energy Transfer by

Resonance", *J. Phys. Soc. Jpn.*, vol. 22, 1967, pp. 779-785.

132. I. M. Rozman, "Theory of Luminescence Quenching in Solutions", *Opt. and Spectrosc*, vol. 4, 1958, Rus. p. 536.

133. L. D. Zusman, *Broadening and Quenching of Luminescence in Solutions*, Dissertation for Doctorate Degree in Phys. Math., Novosibirsk (in Russian).

134. A. I. Burshtein and L. D. Zusman, "Concentration Quenching in the Case of Exchange Energy Losses", *Opt. and Spectrosc*, vol. 38, 1975, Rus. p. 1020.

135. J. Heber, "Energy Transfer and Rate Equations - Application to Ruby", *Phys. Status Solidi*, vol. 42, 1970, pp. 497-506.

136. D. L. Huber, "Donor Fluorescence at High Trap Concentration", *Phys. Rev. B. Solid State*, vol. 20, 1979, pp. 5333-5338.

137. N. D. Zhevandrov, "Polarized Luminescence of Molecular Crystals", *Trudy FIAN SSSR*, vol. 25, 1964, p. 3 (in Russian).

138. N. D. Zhevandrov and T. V. Il'inih, "Investigation of Energy Migration Between Doped Molecules in the Lattice of Naphthalene by the Polarization Technique", *Bull. Acad. Sci. USSR, Phys. Ser.*, vol. 37, 1973, Rus. p. 474.

139. V. I. Nikitin, M. S. Soskin, and A. I. Khizhnyak, "New Results on the 1.06 μm Luminescent Band Structure of Nd^{3+} in Silicate Glass", *Sov. Tech. Phys. Lett.*, vol. 2, 1976, p. 64.

140. V. I. Nikitin, M. S. Soskin, and A. I. Khizhnyak, "Uncorrelated Inhomogeneous Broadening: the Principal Cause of Narrowband Lasing by Phosphate Glass Containing Nd^{3+} Ions", *Ibid.*, vol. 3, 1977, p. 5.

141. V. I. Nikitin, M. S. Soskin, and A. I. Khizhnyak, "Effect of Uncorrelated Inhomogeneous Broadening of the Nd^{3+} 1.06 μ band on the Laser Properties of Neodymium Glass", *Sov. J. Quant. Electr.*, vol. 8, 1978, p. 788.

142. V. I. Nikitin, M. . Soskin, and A. I. Khizhnyak, "Deformation of the 1.06 μ Nd^{3+} Band Profile of Glasses Under Free-Oscillation Conditions", *Ibid.*, vol. 9, 1979, p. 1314.

143. A. K. Przhevuskii, V. A. Savost'yanov, and M. N. Tolstoi, "Chronospectroscopic Investigation of the Luminescence of Neodymium Glasses", *Ibid.*, vol. 8, 1978, p. 54.

144. V. A. Savost'yanov, V. A. Malyshev, A. K. Przhevuskii, and A. S. Troshin, "Chronospectroscopic Study of the Luminescence-Band Structure for Yb^{3+} Ions in Glass at Low Temperature", *Opt. and Spectrosc*, vol. 47, 1979, p. 295.

145. L. E. Ageeva, A. K. Przhevuskii, M. N. Tolstoi, and V. N. Shapovalov, "Effect of Energy Migration on the Position of the Luminescence Band of a Rare-Earth Activator in Glass", *Sov. Phys. Solid State*, vol. 16, 1974, Rus. p. 1659.

146. E. N. Alekseev, V. P. Gapontsev, M. E. Zhabotinskii, and Yu. E. Sverchkov, "Measurement of Nonresonant Interactions", *Sov. JETP Lett.*, vol. 27, 1978, p. 109.

147. Yu. V. Denisov and V. A. Kizel, "Energy Migration in Europium-Activated Borate Glasses and Relative Location of Energy Levels", *Opt. and Spectr.*, vol. 23, 1967, p. 251.

148. A. K. Przhevuskii, "Excitation Migration Through Active Centers in Crystals and Glasses", *Bull. Acad. Sci. USSR, Phys. Ser.* vol. 45, 1981, p. 4.

149. V. A. Savostyanov and A. K. Przhevuskii, "Energy Gap Dependence of Excitation Migration Along a Uniform Contour in Glass", *Sov. Phys. Solid State*, vol. 21, 1979, p. 465.

150. O. K. Alimov, T. T. Basiev, E. O. Kirpichenkova, and S. B. Mirov, "Kinetic Luminescent Spectroscopy of Nd^{3+}, Yb^{3+} and Cr^{3+} Ions in Solids Under Selective Laser Excitation", In: *Abstracts of All-Union Vavilov Symposium on Luminescence*, Leningrad: GOI, 1981, p. 30 (in Russian).

151. O. K. Alimov, M. Kh. Ashurov, T. T. Basiev, and Yu. K. Voron'ko, *Selective Excitation Spectroscopy of Yb^{3+} and Nd^{3+} Ions in Phosphate Glasses and Ion-Ion Interactions*, Preprint of the Lebedev Physics Institute, No. 77, Moscow (in Russian).

152. T. T. Basiev, "Tunable Color-Center Lasers and Their Use in Selective Spectroscopy of Disordered Media", *Bull. Acad. Sci. USSR, Phys. Ser.*, vol. 49, 1985, p. 68.

153. T. T. Basiev, "Selective Kinetic Spectroscopy of Quenching and Migration of Optical Excitation in Disordered Solids", *J. Phys.* (France), vol. 46, 1985, pp. C7-159 - C7-163.

154. A. S. Agabekyan, "Theory of Resonant Energy Transfer Accounting for Inhomogeneous Spectral Line Broadening", *Opt. and Spectrosc.*, vol. 30, 1971, Rus. p. 449.

155. T. Holstein, S. K. Lyo, and R. Orbach, "Spectral-Spatial Diffusion in Inhomogeneously Broadened Systems", *Phys. Rev. B. Solid State.*, vol. 15, 1977, pp. 4693-4698.

156. S. K. Lyo, "Study of Time-Dependent Spectral Transfer Disordered Systems", *Ibid.*, vol. 20, 1979, pp. 1297-1303.

157. V. P. Gapontsev, F. S. Dzheparov, N. S. Platonov, and V. E. Shestopal, "The Kinetics of Excitation Delocalization in Disordered Systems", *Sov. Phys. Lett.*, vol. 41, no. 11, 1985, p. 561.

158. M. J. Weber, J. A. Paisner, S. S. Sussman, et al., "Spectroscopic Studies of Rare-Earth Ions in Glass Using Fluorescence Line-Narrowing Technique", *J. Luminescence*, no. 12/13, 1976, pp. 729-735.

159. P. Avouris, A. Compion, and M. A. El-Sayed, "Phonon Assisted Site-to-Site Electronic Energy Transfer Between Eu^{3+} Ions in an Amor-

202

phous Solid", *Chem. Phys. Lett.*, vol. 50, 1977, p. 9-13.

160. M. A. El-Sayed, A. Compion, and P. Avouris, "Temperature Temporal and Concentration Dependence of the Laser-Narrowed $^5D_0-^7F_0$ Fluorescence Line-Shape of Eu^{3+} in Glasses", *J. Mol. Struct.*, vol. 46, 1978, pp. 355-361.

161. M. Kh. Ashurov, T. T. Basiev, A. I. Burshtein, et al., "Diffusive Delocalization of Electronic Excitations Through a Disordered System of Centers", *Sov. JETP Lett.*, vol. 40, 1984, p. 841.

162. M. Kh. Ashurov, T. T. Basiev, A. I. Burshtein, et al., "Investigation of Diffusive Delocalization of Electronic Excitations Through a Disordered System of Centers by $LiF(F_2 - F_2^+)$ Color Center Lasers", *Proc. IY All-Union Conf. "Tunable Lasers"*, Novosibirsk: ITF SO AN SSSR, 1983, p. 130 (in Russian).

163. L. Gomez-Jahn, J. Kasinsky, and R. J. Dwayne Miller, "Spatial Properties of Energy Transport in Three-Dimensional Disordered Systems", *J. Phs.* France, C, vol. 46, 1985, pp. 85-90.

164. V. P. Lebedev, A. K. Przhevuskii, "Polarized Luminescence of Glasses Activated by Rare-Earth Ions", *Sov. Phys. Solid State*, vol. 19, 1977, Rus. p. 1373.

165. A. K. Przhevuskii, "Inhomogeneous Spectral Structure of Glasses Doped with Rare-Earth Ions", In: *Spectroscopy of Crystals*, Ed. by P. P. Feofilov, Leningrad: Nauka, 1978, p. 96 (in Russian).

166. V. P. Lebedev, A. K. Przhevuskii, "Determination of the Multipolarities of Optical Transitions in the Spectra of Glasses Activated with Eu^{3+} Ions by Means of Polarized Luminescence", *Opt. and Spectrosc.* vol. 48, 1980, Rus. p. 932.

167. T. T. Basiev, M. A. Borik, Yu. K. Voron'ko, et al., "Selective Laser Excitation of Luminescence of Sm^{3+} Ions in Lanthanoalumosilicate Glass", *Opt. and Spectrosc.* vol. 46, 1979, p. 510.

168. A. G. Avanesov, T. T. Basiev, Yu. K. Voron'ko, et al., "Selective Laser Excitation Study of Electron-Phonon Interaction of Sm^{3+} Ions in Glass", *Ibid.*, vol. 51, 1981, Rus. p. 148.

169. K. K. Palkina, V. G. Kuznetsov, N. N. Chudinova, and N. T. Chibisova, "Crystalline Structure of $KNd(PO_3)_4$"; *Proc. Acad. Sci. USSR*, vol. 226, 1976, Rus. p. 357.

170. J. D. Dow, "Resonance Energy Transfer in Condensed Media from a Many-Particle Viewpoint", *Phys. Rev.*, vol. 174, 1968, pp. 962-976.

171. T. T. Basi8ev, Yu. K. Boron'ko, S. B. Mirov, and A. M. Prokhorov, "Frequency Selection of Nd^{3+} Ions in Glass, Excited by Monochromatic Laser Radiation at the $^4I_{9/2}-^4F_{3/2}$ Resonant Transition", *Sov. JETP Lett.*, vol. 29, 1979, p. 639.

172. V. B. Anzin, T. T. Basiev, Yu. K. Voron'ko, and Yu. V. Kosichkin, "Deceleration of Energy Migration Through a $(Y_{0.46}Yb_{0.54})_3Al_5O_{12}$

Crystal in an External Magnetic Field", *Abstracts of the VII All-Union Sympos. on the Spectroscopy of Crystals, Activated Rare-Earth and Transition Metals*, Leningrad: LIYaF, 1982, p. 185 (in Russian).

SUBJECT INDEX